IMAGES OF WAR
M1 ABRAMS TANK

RARE PHOTOGRAPHS FROM WARTIME ARCHIVES

Michael Green

Pen & Sword
MILITARY

First published in Great Britain in 2015 by
PEN & SWORD MILITARY
An imprint of
Pen & Sword Books Ltd
47 Church Street
Barnsley
South Yorkshire
S70 2AS

Copyright © Michael Green, 2015

ISBN 978-1-47383-423-1

The right of Michael Green to be identified as author of this work has been asserted by him in accordance with the Copyright, Designs and Patents Act 1988.

A CIP catalogue record for this book is available from the British Library.

All rights reserved. No part of this book may be reproduced or transmitted in any form or by any means, electronic or mechanical including photocopying, recording or by any information storage and retrieval system, without permission from the Publisher in writing.

Typeset by Concept, Huddersfield, West Yorkshire HD4 5JL.
Printed and bound in India by Replika Press Pvt. Ltd.

Pen & Sword Books Ltd incorporates the imprints of Pen & Sword Archaeology, Atlas, Aviation, Battleground, Discovery, Family History, History, Maritime, Military, Naval, Politics, Railways, Select, Social History, Transport, True Crime, and Claymore Press, Frontline Books, Leo Cooper, Praetorian Press, Remember When, Seaforth Publishing and Wharncliffe.

For a complete list of Pen & Sword titles please contact
PEN & SWORD BOOKS LIMITED
47 Church Street, Barnsley, South Yorkshire, S70 2AS, England
E-mail: enquiries@pen-and-sword.co.uk
Website: www.pen-and-sword.co.uk

Contents

Dedication **4**
Acknowledgements **5**
Foreword **6**
Notes to the Reader **8**

Chapter One
    Abrams Background History **9**

Chapter Two
    M1 through M1A1 Abrams Tanks **45**

Chapter Three
    M1A2 Abrams Tanks **94**

Chapter Four
    Marine Corps Abrams Tanks **121**

Chapter Five
    Firepower Close-Up **137**

Chapter Six
    Specialized Vehicles and Accessory Kits **160**

Dedication

I would like to dedicate this book to the late Dr Philip W. Lett, Jr. considered to be the father of the M1 Abrams tank. He was the head of the Chrysler Corporation Defense Division design and development team for the vehicle, which later became part of General Dynamics Land Systems Division. Prior to working on the M1 Abrams tank he played an important role in the design and development of both the M48 and M60 series of tanks.

Acknowledgments

As with any published work, authors must depend on a great many people for assistance. These included over many years, my fellow author, and my long-time mentor, the late Richard Hunnicutt. Others who greatly assisted me in acquiring the research information and photographs needed for this work include Randy Talbot, the Command historian for the TACOM Life Cycle Management Command (LCMC), and Peter Keating, director of communications for the Land Systems Division of General Dynamics (GDLS).

The US Army Brotherhood of Tankers (USABOT) also provided valuable assistance in the preparation of this work. The organization is committed to preserving the history of US Army tankers and their tanks. Those interested in joining USABOT can find the information needed on their website (www.usabot.org). Additional pictures were acquired from the Defense Video & Imagery Distribution System (DVIDS). For the sake of brevity, it will be listed as DOD in the photo credits.

Foreword

The mantra of the tanker is to move, shoot, and communicate. No better tank was ever created to do this than the M1 Abrams Main Battle Tank. In its various stages of evolution over the past four decades, no vehicle in the history of land warfare has been able to accomplish these three core requirements better than this tank. Not only is it fast and highly mobile, but it can destroy targets with a first round hit thousands of meters away in day or night while on the move, and can communicate by providing key battlefield information to the crew regardless of the chaos of battle.

When I was first introduced to the M1 as a young soldier it was like nothing I had seen before. Sleek, fast, and with the latest technology, it was a vision of the future. I would go on to serve on other armored vehicles before I was finally able to command an M1A1 in 1988 while serving as the commander of Delta Company, 3rd Battalion, 35th Armor, 1st Armored Division in Germany.

The Cold War was beginning to thaw at that time, although there was plenty of uncertainty in the minds of soldiers serving in Europe. The training was still rigorous with semi-annual tank gunnery exercises where we fired hundreds of main gun rounds and thousands of machine gun bullets against tough standards. Field training was equally demanding with the creation of the Combat Maneuver Training Center (CMTC) at Hohenfels, Germany, nicknamed 'NTC-East' in homage to the National Training Center at Fort Irwin, California. And of course there were frequent alerts, deployments to local assembly areas, and exercises in our general defensive positions.

Had the Cold War gone hot, the tankers serving on the front lines of Europe would have deployed to their battle positions in the best tank in the world. The armored protection, power of the main gun, and Nuclear-Biological-Chemical protection it provided would have allowed the crews to not only engage and destroy enemy forces, but to live to fight another day.

Many of these same highly trained M1A1 crews who witnessed the fall of the Berlin Wall and the end of the Warsaw Pact found themselves facing a new enemy not even identified in early 1990. When US forces engaged the Iraqi Army in February of 1991, the performance of the M1A1 tank proved that it was a magnificent fighting machine.

When the Abrams rolled in Bosnia a few years later, the opposing forces knew immediately that this was not a vehicle to tangle with. And, when US forces rolled again into Iraq in 2003, it was Army tank battalions that led 'Thunder Runs' into the heart of Baghdad to formalize the end of the Hussein regime. This most battle tested tank of modern warfare had made its mark in history.

Author Michael Green has produced an exceptional work on the M1 series tank that is easy to follow and understand. Green captures how the Army has upgraded and adapted the Abrams to meet the changing combat environment it must operate in; from the plains of Central Europe, to battles fought in the deserts of the Middle East, and in large urban environments. His easy to follow narrative and photos make this a read to be enjoyed by veteran tankers, history buffs, and those who simply want to learn about the history, and future, of this great tank.

Colonel (retired) John D. Blumenson, US Army Reserve

Notes to the Reader

1. The US Army implemented a new system of tank nomenclature in 1950. No longer were tanks divided by weight into light, medium, or heavy tanks, as had been the practice since the First World War. Rather, they were now classified by the caliber of their main guns. Hence, you have the official US Army Ordnance Department designation of Tank, Combat, Full Tracked, 105mm Gun, M1 General Abrams. For the sake of brevity and readability the author will condense these official designations.
2. The letter 'M' in a vehicle designation code meant it had been standardized (accepted into full service). Additional letters and numbers following the original designation represent improved models of the original vehicle.
3. Due to the fact that in most cases the photographers who took many of the pictures for this work were not aware of what model Abrams tank they were looking at, only when the photographer knew what model tank they shot will the author indicate the vehicle's acronym in a caption, or when some external vehicle or background feature easily marks the version of the tank pictured and where it might be located.
4. The writer of this book had no access to classified information in the preparation of this work. Everything included in the text can be found in open sources, typically online, such as manuals for the various versions of the tank. Additional information on the Abrams tank series can be seen in various YouTube videos, or read in previously published books and articles on the vehicle.

Chapter One

Background History

The US Army Armor Branch and the inventory of tanks that it employed during the Second World War had not earned a sterling reputation, despite the bravery of the tankers that had engaged the Axis with them. US Army General George S. Patton, Jr., who had commanded the famous Third Army during its tank-led advance across Western Europe between July 1944 and May 1945, admitted upon the conclusion of the Second World War: 'I don't have to tell you who won the war, you know our artillery did.'

The blame for the poor wartime performance of American tanks cannot be laid on the engineers that had been tasked with their design and development. The vehicles they had been asked to come up with were shaped by a doctrine envisioned by the US Army's senior leadership of that time in which tanks were weapons of deep attack (exploitation) and infantry support systems, not meant as tank-versus-tank combat vehicles. The latter role was to be performed by specially-designed vehicles referred to as 'tank destroyers.' Combat would show that the doctrine of tank destroyers was badly flawed.

It took until the closing months of the war in Europe before the US Army tankers finally received a small number of a vehicle optimized for the destruction of enemy tanks. That vehicle was the M26 Pershing tank, originally labeled as a heavy tank, and then postwar as a medium tank. It was armed with a 90mm main gun, derived from an anti-aircraft gun design. Postwar there was a slightly improved M26A1 Pershing medium tank fielded.

With the Empire of Japan finally conceding defeat in September 1945, prompted by the American military use of the atomic bombs in August 1945, the civilian leadership of the American government believed that with its monopoly on nuclear weapons the chances of future land wars had greatly decreased. They therefore saw no reason to fund the next generation of postwar tanks. The US Army was therefore pushed into modernizing its existing tank fleet, beginning in 1948, with what money was available. This resulted in the conversion of the bulk of its fleet of approximately 2,200 M26/M26A1 Pershing medium tanks into the improved M46 Patton medium tank, as a stopgap measure.

Game Changers

The belief by the civilian leadership of the American government of the unlikely occurrence of any more land wars began changing with two major events. These included the Soviet Union exploding its first atomic bomb in August 1949, which brought a quick end to the country's monopoly of nuclear weapons. That was followed in the summer of 1950, when North Korea, a communist client state, attacked South Korea, an American client state. These became the opening acts in what historians refer to as the Cold War, which lasted from 1946 until 1991.

The North Korean Army spearheaded its invasion of South Korea in the summer of 1950 with approximately 150 Soviet-supplied T34-85 medium tanks, which quickly brushed aside the American trained and equipped South Korean Army, as well as US Army units armed with the M24 Chaffee light tank sent to assist them. This dramatic turn of events convinced all concerned that the same thing could happen in Western Europe, but on a much larger scale, if the tank-heavy Soviet Army or client state armies assembled in Eastern Europe were to attack.

The American Response

As a countermove to the success the North Korean tanks enjoyed early on in the Korean War (1950–53), the American Congress opened its purse strings and authorized the funding of the next generation of tanks for the US Army. They would be in theory superior to their more numerous Soviet-designed and built counterparts they would encounter in the opening stages of a Third World War in Western Europe.

Unfortunately, due to the long lead time involved in the design and development of a new tank, the US Army was forced to make do with another stopgap vehicle. It consisted of the turret from an experimental medium tank fitted with a new and advanced fire-control system mounted on the chassis of the existing M46 Patton medium tank. This combination of tank components was standardized in 1952 as the M47 Patton medium tank. It was rushed into production for service in Western Europe, without the normal testing process being completed beforehand.

The speed with which the M47 tank was ordered into production by the US Army, and the lack of testing, not surprisingly led to an almost unending series of teething problems that would seriously compromise the tank's effectiveness throughout its service life. There were 8,576 units of the M47 tank built, with production ending in 1953. Having quickly soured on the M47 tank, the US Army provided the majority of them to its new NATO allies in the 1950s in an attempt to standardize weapons systems. The US Army now pinned all its hopes on its replacement, the M48 Patton medium tank, standardized in 1953.

Last of the Patton Tanks

Despite the numerous design problems encountered by rushing the M47 tank into service, the US Army was still under the perceived threat that at any minute the

Soviet Army, and its untold thousands of tanks, would cross from Eastern Europe into Western Europe, and it decided to rush the M48 tank into service without any prior testing. As with the M47 tank, this decision proved to be a serious mistake. The first production version of the M48 tank was deemed by the American Government Accounting Office (GAO), now known as the Government Accountability Office, as not even being fit as a training vehicle.

By 1959, when production of the M48 tank series had ended, approximately 12,000 units in various versions had been constructed. It took until 1963, with the introduction of an upgraded model of the vehicle labeled the M48A3 tank, for the design to mature and the machine to become a reliable workhorse. By this time however, the tank's 90mm main gun was obsolete against many of the Soviet tanks then in service, the most common being the T-54 medium tanks, armed with a 100mm main gun.

The T-54 medium tank began entering widespread service with the Soviet Army and many of its client states in the early 1950s. It was complemented, beginning in 1958, by an upgraded version designated the T-55 tank. Production of the T-54 tank series, including license-built copies, continued until 1959. Production of the T-55 tank series lasted until 1981. The approximate number of T-54/T-55 series tanks is estimated anywhere between 86,000 to 100,000 units, making them the most-produced tanks in history.

The American Answer to the T-54/T-55 Medium Tank

In order to field a tank quickly, and in large numbers, armed with a main gun large enough to penetrate the frontal armor array of the Soviet T-54/T-55 tank series, the US Army took a single M48A2 tank in very early 1959 as an experiment and armed it with an American version of a brand new British-developed 105mm tank gun, designated the L7. The British Army had adopted the weapon in 1959 for mounting in their Centurion Mark 7 tank. The modified American-built copy of the tank gun was designated the M68.

The design and development of the L7 105mm British tank gun had been initiated by a chance event during the 1956 Hungarian Uprising. Somebody drove a captured Soviet Army T-54 tank onto the grounds of the British Embassy, located in the Hungarian capital of Budapest. Before returning the tank to the Soviet Army, the British military experts on staff closely examined the vehicle and realized that their existing main gun on the Centurion tank series, the 20-pounder (83mm), lacked the penetrative power to deal with it. Therefore a new larger and more powerful tank main gun was required, resulting in the quick development and fielding of the L7 105mm tank gun.

In addition to the new 105mm M68 main gun, the single modified M48A2 tank was fitted with a better-protected front hull configuration. Another big improvement was the replacement of the original gasoline-powered engine, with a much more reliable and fuel-efficient diesel-powered engine. Testing of this single modified M48A2 tank went very well and in March 1959 the US Army re-designated it the M60 tank, and ordered it into full-scale production, as a replacement for the M48 tank series. It was at the time considered only an interim vehicle.

The M60 was not the original choice of the US Army to replace the M48 tank series. Late 1954, Chief of Staff of the US Army approved the development of a new medium tank under the T95 tank program. Despite many innovative design features, the program proved to be a dead end, and the US Army lost interest in it by 1958.

Between 1959 and 1982, over 15,000 units of the M60 tank would be built in four models. An upgraded version with an elongated turret offering superior ballistic protection was the M60A1 tank, and entered service in 1961. The M60A2 tank was armed with a 152mm gun/Shillelagh missile launcher instead of a 105mm main gun.

Despite being a progressively improved version of the M48 tank series, which meant almost all the teething problems often associated with new tanks were absent from the M60 tank series, it was not considered by the US Army as part of the Patton tank series (M46 through M48). Neither was it ever assigned an official nickname by the US Army, despite some unofficial nicknames being applied to it by those who served on them; such as 'Super Patton'.

The M60 series tanks were essentially medium tanks equipped with a main gun that was the equal in performance of the 120mm main gun fitted in the M103 series heavy tank. Reflecting the medium and heavy tank roles it took over, the M60 series tanks were often referred to as a 'main battle tank' (MBT). This was a label that first appeared in US Army documents in 1957. The Centurion Mark 7, with the 105mm main gun, performed the same role as the new M60 series tanks for the British Army, resulting in the discarding of their postwar heavy tank, the Conqueror. Instead of the term main battle tank, the British Army adopted the term 'universal tank'.

The Soviet Answer to the M60 Tank

The Soviet Army became concerned in the 1950s that the frontal armor array on the American M48 tank series, and its British counterpart, the Centurion tank series, might be proof against the tank-killing rounds fired from the 100mm main guns mounted in their T-54 and T-55 series tanks. That concern grew even larger when a brand new M60A1 tank came into their possession in January 1961, as a turncoat Royal Iranian Army officer drove a newly issued example across the Iranian border into the Soviet Union. Upon closely examining the new American tank they realized that its gun and armor made it far superior to their existing T-54/T-54 series tanks.

To deal with the M60A1 tank and penetrate its frontal armor array, the Soviet Army set about the design and development of a more capable 115mm tank main gun. Instead of being rifled as all tank main guns had been since the beginning, the new

Soviet-designed and built 115mm tank gun was smooth bore. Because the new 115mm tank gun was too large to fit within the existing T-54/T-55 series tank turrets, it was decided to develop a new larger turret that would be mounted on a lengthened T-55 tank chassis. The new tank was designated the T-62 medium tank and was accepted for service in 1961, with large-scale production beginning the following year.

From the beginning, the T-62 tank was intended only as a stopgap vehicle to counter the fielding of the American M60 tank series in Western Europe. Despite being only an interim vehicle, the T-62 tank would go on to become the frontline vehicle for the Soviet Army into the 1970s, with approximately 20,000 units built by the time production ended in 1966. The majority of the T-62 tanks would be based in Eastern Europe and tasked with the invasion of Western Europe, upon the outbreak of a Third World War.

The US Army Looks to the Future

Even as the initial models of the M60 tank series began entering service in the early 1960s, the US Army was thinking about their next generation MBT. It would be intended to deal with whatever new tank or tanks the Soviet Army might be planning for the future. These would eventually include the T-72, T-64, and T-80 series MBTs. All were armed with a 125mm smoothbore main gun, which was originally developed to counter the fielding of the British Army Chieftain MBT that entered service in 1967.

The reason the US Army began thinking about a new MBT so soon after fielding the M60 MBT was the awareness that the design and development cycle of ever more complex tanks would require a longer gestation period than the relatively crude and simple tanks of the Second World War. In the postwar era it was generally accepted that it would take at least ten years from concept design to fielding a new MBT.

To provide US Army tankers in Western Europe a marked advantage upon engaging the much more numerous tanks of the Soviet Army in a possible Third World War scenario, the US Army wanted a revolutionary MBT design that would offer a qualitative advantage in battle to offset the numerical advantage in tanks enjoyed by the Soviet Army. What the US Army did not want was an evolutionary development of the M60 tank series, which they felt could no longer offer the qualitative advantage they sought, no matter what upgrades were added.

The Replacement for the M60 Tank Series

As the US Army began testing a number of experimental tank concepts in the 1960s it found itself being pushed by the then American Secretary of Defense (a civilian position) into working together with the West German Army to design a new main battle tank that would meet both their needs. The result of this 1963 agreement was the main battle tank-1970 (MBT70).

Rising costs, a host of design issues, and the US Army turning all its resources to trying to win the Vietnam War resulted in the cancellation of the MBT70 in January 1970. Not wanting to completely give up on its search for a next-generation MBT, the US Army convinced the American Congress to fund the development of a simplified and hopefully more affordable version of the vehicle eventually known as the XM803. Unfortunately, the US Army failed to show that the XM803 would be any less costly to put into production than the MBT70 and the American Congress cancelled it in December 1970.

The Third Time is the Charm

Having learned some painful lessons from the demise of the MBT70 and the XM803, the US Army worked very hard to get it right on the next go-around. Because of some controversy within the US Army regarding what form a new MBT should take, a group of leading US Army armor experts was brought together in February 1972 and labeled the MBT Task Force (MBTTF). Their job was to define the characteristics required of a next generation MBT.

A pressing requirement of the MBTTF was to keep the vehicle's cost down, so as not to alarm the American Congress, which might then pull the funding for the proposed MBT (later designated the XM1). This matter is addressed in a US Army Center for Military History Publication titled *Developing the Armored Force: Experiences and Visions*, an interview with US Army Major General Robert J. Sunell (retired). In it, one of the interviewers asks Sunell about the cost limit imposed on the proposed XM1 tank. He answered:

> General Baer [tank program manager], in effect, said, 'This is what all these different items cost on a tank, and I have $507,000 ceiling for the XM1 tank.' And he said, 'If you really want to add that to the tank, here's what a fender costs; here's what a machine gun costs. Now which one do you want me to take off, because I cannot exceed this cost ceiling?' Everybody understood that.

To assist the MBTTF, contracts were awarded by the US Army to General Motors and the Chrysler Corporation to begin advance design work on the proposed new MBT, then labeled the XM815. In May 1973, the two civilian contractors were awarded contracts to build three prototypes each of their proposed new MBTs, which would reflect the findings of the MBTTF. They were to be delivered by early 1976 for extensive testing. General Motors went for a diesel-engine powered tank, and Chrysler a gas-turbine engine powered tank.

A New Type of Armor Appears

In July 1973, the US Army invited selected representatives from both General Motors and Chrysler to the United Kingdom to see a newly-invented armor unofficially named 'Chobham,' after the location at which it was invented. It offered a superior level of protection from the ever-growing threat of shaped charge warheads that had grown in numbers and effectiveness since the Second World War. The official British name for the new product was 'Burlington' armor, which has been described in the open Press as a multi-layered composite of different materials. It has now been declassified and information on its construction posted online.

The problem with Burlington armor was that in order for it to be effective on the XM815 the vehicle's weight would rise to at least 58 tons. Some in the MBTTF were strongly against fielding a new MBT that heavy, and preferred keeping the vehicle at no more than 45 tons. The decision to go with an American version of Burlington armor for the XM815, and accept the weight penalty, was made by the then Chief of Staff of the US Army, General Creighton Abrams, in September 1973.

Rather than go with the British-invented and built Burlington armor for the XM815, the US Army went for a modified American-built version referred to as 'special armor'. General Motors and Chrysler quickly re-designed their proposed MBTs to incorporate the new American-built special armor within their tank's turrets and hulls.

The Companies Present Their Products

As called for by the US Army contracts with General Motors and Chrysler, both firms had their three prototypes of their respective versions of the new MBT ready for testing in early 1976. The XM815 had been relabeled the XM1 in late 1973. By this time, the General Motors prototypes had been re-powered with gas-turbine engines, as the US Army had decided that gas-turbine engines offered more performance and reliability than existing diesel engines.

The testing of the two competing companies' prototypes continued until May 1976. At one point early during the testing cycle, there was a delay as the idea of the US Army and West German Army joining together to come up with a new MBT was reintroduced. Thankfully, that idea soon fell to the wayside, other than an open agreement that at a certain point in the future, the XM1 would be armed with a German-designed 120mm main gun rather than the British-designed 105mm main gun it began its service life with.

A Winner is Chosen

On completion of the testing between the General Motors and Chrysler prototypes of the XM1, it was announced by the US Army in November 1976 that Chrysler had won the competition. Some in the US Army armor community preferred the General Motors prototype. They believed that the awarding of the contract to Chrysler was a political decision intended to provide that badly struggling corporation with a financial lifeline to keep it from bankruptcy and the resulting large number of workers becoming unemployed.

Upon winning the XM1 tank contract Chrysler was required to supply eleven pilot tanks for additional testing, which would help in determining the final design for a batch of low-rate initial production (LRIP) vehicles. Pilot tanks are typically the last stage before the building of series production vehicles.

Chrysler delivered the first of the 11 LRIP vehicles in February 1978 and the last in July 1978. The Secretary of Defense approved the construction of an additional 110 LRIP vehicles in May 1979. On 28 February 1980 the US Army had the official rollout ceremony for the XM1, and it was at this point that they announced that the tank was to be named the 'Abrams' in honor of General Creighton W. Abrams, who had died of cancer in September 1974, at the age of 59.

Abrams had commanded the 37th Tank Battalion, 4th Armored Division, during the Second World War. Among his many battlefield accomplishments was the relief of the besieged village of Bastogne, in which the 101st Airborne Division, as well as other US Army units, had been surrounded and cut-off by numerous German divisions during what is known as The Battle of the Bulge. General Patton would say of Abrams near the end of the Second World War, 'I'm supposed to be the best tank commander in the army but I have one peer, Abe Abrams. He's the world champion.'

The LRIP vehicles went straight from the Chrysler production line into testing, which began in September 1980 and was to last through May 1981. As with its earlier postwar tank designs, such as the M47 and M48 medium tanks, the US Army felt it had to deploy the XM1 Abrams tank to Western Europe as soon as possible to counter the ever present Soviet Army threat to that area of the world. It was therefore willing to rush the vehicle into production before its testing was complete and risk the consequences. Fortunately, once the XM1 overcame some initial teething issues it became an extremely reliable tank in very quick order, unlike the earlier models of the Patton medium tank series.

Series Production is Approved

In February 1981, with testing still going on, the US Army ordered series production of the XM1 Abrams tank, to begin at an initial rate of thirty vehicles per month. It was later raised to sixty vehicles per month in November 1981. That same month, the US Army standardized the XM1 Abrams tank and therefore dropped the letter 'X' from its designation code. The vehicle therefore became the M1 Abrams tank, although American tankers often simply referred to it as the 'M1', or just the 'Abrams'. On the front of the US Army manuals for the tank it is listed as Tank, Combat, Full Tracked, 105mm Gun, M1, General Abrams.

Due to financial problems with its car divisions, Chrysler was forced to sell its military product division, Chrysler Defense Corporation, in 1982. The American government quickly approved its sale to the General Dynamics Corporation, which named its newest acquisition General Dynamics Land System (GDLS).

The mainstay of the US Army's tank fleet during the Second World War was the M4 medium tank. Pictured is a restored M4A1 model, armed with a 75mm main gun. An adequate vehicle when introduced into service in 1942, it was clearly deficient by the time the US Army invaded France in the summer of 1944, in both firepower and armor protection. *(Michael Green)*

A restored M4A3 medium tank armed with a 75mm main gun. The M4A3 hull was constructed of steel plates welded together instead of the cast armor of the M4A1 medium tank. The type of armor used on the M4 medium tanks meant little as all could be easily penetrated by the entire range of German antitank weapons. *(Michael Green)*

(*Above*) Even before the first generation of the M4 medium tanks began rolling off the assembly line, the US Army Ordnance Department began looking down the line at a second generation of improved up-armored and up-gunned M4 medium tanks. A restored example of that line of development is seen here armed with a 76mm main gun. The particular version pictured was designated the M4A3(76)WHVSS. (*Michael Green*)

(*Opposite above*) To deal with enemy tanks the US Army had decided to deploy specialized lightly-armored full-tracked vehicles armed with main guns of sufficient caliber to deal with whatever they encountered. An example of that doctrine was the M10 Tank Destroyer seen here armed with a 3-inch main gun. Unfortunately, the vehicle's main gun could not penetrate the frontal armor array of late-war German tanks, or self-propelled guns. (*Michael Green*)

(*Opposite below*) Another attempt to field a tank destroyer with a sufficient degree of firepower to deal with late-war tanks and self-propelled guns was the M18 tank destroyer. Pictured is a restored example of the vehicle. Rushed into service, the M18 was plagued early on by a number of design issues. The most serious issue was the fact that its 76mm main gun was not up to penetrating the frontal armor array on enemy tanks, and self-propelled guns. (*Pierre-Olivier Buan*)

(Opposite above) The restored M36 Tank Destroyer seen here was the intended replacement for the M10 Tank Destroyer. The M36 appeared in Western Europe in late 1944 and was armed with a 90mm main gun that at certain ranges could penetrate the frontal armor array of late-war German tanks and self-propelled guns. Like the other tank destroyers that came before, it proved badly under-armored for the battlefield. *(Pierre-Olivier Buan)*

(Above) The M24 light tank pictured was a late-war addition to the US Army fighting in Western Europe during the Second World War. It remained in service postwar and was the only tank available to send to South Korea with the outbreak of war in June 1950. Up against North Korean T34-85 medium tanks in July 1950, they did not fare well. *(Michael Green)*

(Opposite below) A last-minute addition to the US Army tank inventory in 1945 was the T26 heavy tank. It was armed with the same 90mm main gun as the M36 Tank Destroyer, but had much thicker armor to survive a wider range of German antitank weapons. After the Second World War the vehicle was relabeled a medium tank and designated the M26 Pershing medium tank; an example is seen here. *(Bob Fleming)*

(Opposite above) The North Korean Army had an inventory of 150 units of the Russian T34-85 medium tank, seen here, when they invaded South Korea. In the early stages of the conflict they completely dominated the opposition. Their time as the king of the battlefield in the Korean War ended when the US Army introduced the M4A3(76)WHVSS, the M26, M26A1, and M46 medium tanks into the conflict. *(Michael Green)*

(Opposite below) The early success of the North Korean T34-85 medium tanks during the Korean War pushed the US Army to seek out a new medium tank to replace the existing M46 medium tank (an upgraded M26A1 medium tank). What eventually rolled off the factory floor in 1950 was the M47 medium tank, seen here, which consisted of an M46 chassis and a new experimental turret. *(Vladimar Yakubov)*

(Above) Unresolved design issues with the M47 medium tank led the US Army to consider it a failure in service. It was therefore superseded by the M48 medium tank in April 1953. The most noticeable external difference between the two tanks was the rounded turret, seen here on an M48 tank, which contrasted sharply with the wedge-shaped turret of the M47 tank. Both tanks were armed with the same 90mm main gun. *(Michael Green)*

(*Opposite above*) The British Army's early Cold War counterpart to the US Army Patton medium tank series was the Centurion tank series. The restored example pictured here belongs to the Tank Museum and is a Mark 3 armed with a QF 20-pounder (84mm) main gun. Unlike the US Army M47 and M48 medium tanks, it did see combat during the Korean War and earned a reputation as a well-armed and well-protected vehicle. (*Tank Museum*)

(*Opposite below*) The M48 Patton medium tank would appear in many models over the years, each an improvement over the last. The final model of the series with the American military was the M48A5 shown here, which entered service between 1975 and 1979. It was armed with a 105mm main gun instead of the 90mm main gun fitted to all earlier models. Most served with the US Army National Guard. (*Michael Green*)

(*Above*) The Israeli Army version of the US Army M48A5 tank seen here at a museum was referred to as the Magach 3. In combat, it completely dominated the Soviet tanks supplied to various Arab armies, including the T-62 medium tank armed with a 115mm main gun, during the 1973 Yom Kippur War. (*Vladimar Yakubov*)

The bulk of the Soviet and Eastern-Bloc tanks employed by the Arab armies in the 1973 Yom Kippur War were various versions of the T-54 and T-55 medium tank series. All were armed with a 100mm main gun. The example pictured is a Czech license-built copy of the Soviet T-55. It is shown in Czech Army markings. (Michael Green)

Armed with an experimental 90mm smooth-bore main gun is one of the US Army's T95 prototypes, showing off its ability to tilt. It could also be raised or lowered with a hydro-pneumatic suspension system. It was later fitted with a tank commander's cupola and an armor-protected searchlight on the right side of the turret. (TACOM)

Pictured is one of the US Army's T95 prototypes having been lowered to its minimum height by way of its hydro-pneumatic suspension system. The US Army believed at the time that a tank with this ability would have a battlefield advantage over opposing tanks that lacked the same feature. The turret on this prototype T95 is experimental and was the forerunner of the M60A2 turret. (TACOM)

With the demise of the T95 tank program, the US Army pushed forward with a new tank that they had begun referring to as a 'main battle tank.' Plans called for it to replace the existing M48 series medium tanks in the inventory. It was also intended to replace the M103 Heavy Tank, armed with a 120mm main gun, as seen here. (TACOM)

(*Opposite above*) To save time the US Army decided to base its new main battle tank, the M60 seen here, on the M48 medium tank chassis. The M60 had a new wedge-shaped glacis, a 105mm main gun, and a diesel engine instead of the gasoline engines used in the M48 medium tanks and M103 heavy tanks. Between 1959 and 1961 a total of 2,205 units of the M60 were constructed. (*Dick Hunnicutt*)

(*Opposite below*) A number of US Army M60A1 tanks are shown during a training maneuver at Fort Bliss, Texas. Hand signals as seen in this picture are still taught to American tankers, but seldom used, especially not in combat. All the tanks visible are equipped with a Xenon searchlight kit which produced one million candlepower and could reach out to 2,000 yards. It could be employed in either white light or infrared. (*DOD*)

(*Above*) The US Army disliked the ballistic protection offered by the large rounded turret of the first model of the M60 main battle tank series, which it had inherited from the M48 medium tank. Therefore a new redesigned elongated-shaped turret, offering a superior level of ballistic protection, was fitted to the vehicle as seen here. This resulted in the vehicle being designated the M60A1. (*TACOM*)

The US Army became enamored with the idea of tanks firing guided missiles in 1959, continuing into the 1960s. They seemed to offer a degree of accuracy at longer ranges that the 105mm main gun on the M60 series tanks then lacked. The result of this interest was the development of the M60A1E1 pilot tank seen here fitted with an experimental turret armed with a 152mm gun/missile launcher. It later became the M60A2 tank. *(TACOM)*

The failure of the M60A2 and the accuracy limitations of the 105mm main gun of the M60A1 tank, combined with the expense of the Vietnam War had drained the US Army's coffers of the funding needed to field a new main battle tank to replace the M60 series. This pushed the service into upgrading the M60A1, which would result in the M60A3 seen here. *(TACOM)*

Pictured in the turret of a US Army M60A3 tank is the vehicle commander. He is holding the tank commander's override control. This feature would allow him to take control of the turret and main gun away from the gunner during an engagement. The tank commander could then point the turret and main gun in the direction of a more serious threat that the gunner needed to engage first. *(DOD)*

The US Army was aware that it needed to stay a step ahead of Soviet tank designs to maintain the battlefield superiority of its own tanks. This resulted in the American government and the West German government coming to an agreement in 1963 to co-produce a tank labeled the main battle tank 1970 (MBT70). A prototype of the MBT70 tank is shown here. *(TACOM)*

(*Above*) A novel feature of the MBT70 prototypes, an example being show here, was the driver being positioned in the turret and not the front hull. To keep the driver facing forward when the turret was rotated, he sat in a small one-man counter-rotating capsule. It was located in the left front of the turret alongside the main gun. There was no loader on the MBT70 as it was fitted with an automatic loader. (*TACOM*)

(*Opposite above*) With the West German government pulling out of the agreement to field the MBT70 in 1969, leaving the US Army to fund the program on its own, the future viability of the program quickly came into question. To nobody's surprise the program was cancelled the following year. To save some of the time and effort put into the MBT70 a simpler and less costly version designated the XM803, seen here, was proposed, but it too was quickly cancelled. (*Patton Museum*)

(*Opposite below*) When the West German government pulled out of the MBT70 program in 1969, they went on to develop a new series of tanks better suited to their own requirements. That series of tanks was labeled the Leopard 2. The first model of the Leopard 2 tank series entered service with the West German Army in 1979. The version of the Leopard 2 tank series pictured is the 'A4' model. (*Krauss-Maffei*)

Despite the earlier missteps, the US Army commenced work on a new tank. Originally it was anticipated that three American car companies would be interested in bidding for the chance to build a new tank for the army. These included Chrysler, General Motors, and Ford. However, Ford declined to enter the competition. Pictured is a Chrysler prototype of the new tank labeled the XM1 firing its main gun. *(TACOM)*

Pictured is one of the Chrysler XM1 prototypes running at high speed. Unlike General Motors, that fitted their XM1 prototypes with diesel engines, Chrysler installed gas turbine engines in their XM-1 prototypes. Chrysler preferred the gas turbine engine because of its smaller size and lighter weight. *(TACOM)*

A General Motors prototype of the XM1 is seen here. It differed externally in a couple of design features from its Chrysler counterpart. One of those features was the gunner's optical sights. It was located in a small armored enclosure projecting out from the right side of the turret. This allowed General Motors to provide the same level of ballistic protection on both sides of the tank's main gun. *(TACOM)*

The front turret armor arrangement on the General Motors XM1 prototypes is seen here. It differed slightly from its Chrysler counterparts. Both firms' vehicles were armed with a 105mm main gun in contrast to the 152mm gun/missile launcher on the MBT70 and XM803. To reduce the cost and complexity of the XM1, neither competitor's prototypes were fitted with hydro-pneumatic suspension systems. *(TACOM)*

The US Army conducted a series of validation tests in 1976 to determine which company's XM1 prototype best met their future requirements. In the end, Chrysler was selected to build the service's new main battle tank. Prior to series production Chrysler was required to construct eleven full-scale engineering development (FSED) or pilot tanks; pictured is a scale model made before they were actually built. *(TACOM)*

Satisfactory testing of the eleven XM1 pilot tanks between 1978 and early 1979 resulted in Chrysler being given the go-ahead to begin low initial rate production (LIRP) of the tank in May 1979. The first XM1 of the 110 LIRP tanks built is seen here on museum display. The name 'Thunderbolt' painted on the turret was the nickname of one of General Creighton Abrams' Second World War M4 medium tanks. *(Christophe Vallier)*

A Chrysler LIRP XM1 tank is shown parked next to an M60A3 tank at Aberdeen Proving Ground, Maryland. Unlike the sleek elongated cast armor turret of the M60A3, the XM1 turret is slab-sided and slightly sloped. This arrangement reflected the fact that the American version of the British-designed Chobham armor employed on the XM1 was only effective in that configuration. *(TACOM)*

Pictured at Aberdeen Proving Ground, Maryland is an LIRP XM1 sailing over a small embankment. High speed was one of the key requirements of the XM1 from the beginning. The US Army believed that a tank that was capable of moving quickly while operating off-road would be much more difficult to track and engage on the battlefield. *(TACOM)*

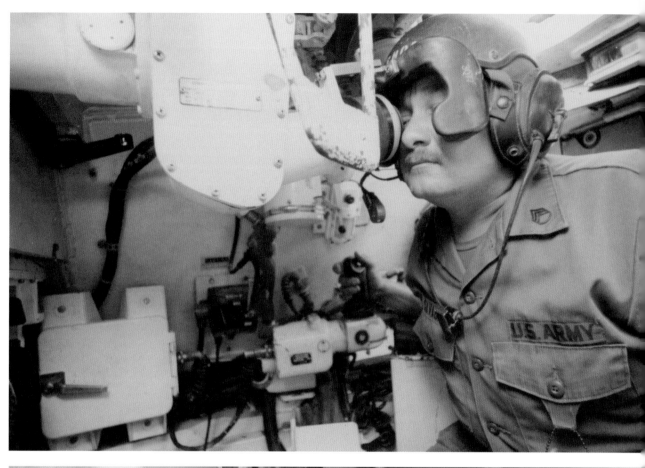

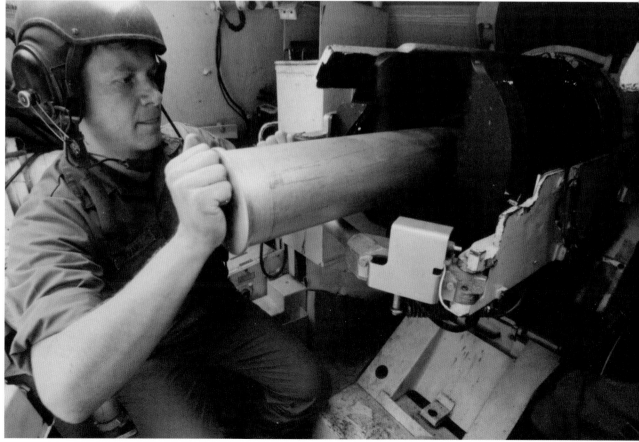

(*Above*) Shown on a firing range is an LIRP XM1. All the variants of the Abrams series of tanks continue to be equipped with a hydraulically driven and electrically controlled main gun turret drive. The main gun turret drive works in conjunction with the tank's stabilization system to provide a stable firing platform when the vehicle is moving. (*TACOM*)

(*Opposite above*) The tank commander on an LIRP XM1 is shown in this photograph looking through the gunner's primary sight (GPS) extension. There is also a gunner's auxiliary sight (GAS), which is an articulated telescope coaxially mounted to the main gun with an eight power magnification and an eight degree field of view. It is the backup if the GPS fails. (*TACOM*)

(*Opposite below*) An LIRP XM1 loader is shown inserting a main gun round into the breech of the tank's 105mm main gun. The portion of the fixed one-piece round visible is referred to as the cartridge case and is made of steel. When the round is fired, the steel cartridge case expands to seal the rear of the gun tube. Upon counter-recoil the steel cartridge case is ejected from the breech. (*TACOM*)

There are a couple of external identifying features for the LIRP XM1. One is the location of the crosswind sensor mounted on a mast at the rear edge of the turret roof between the antenna posts, as pictured. The mast is affixed to a rubber shock mount. Notice the wedge-shaped rear wall of the turret, another external identifying feature of the LIRP XM1. *(GDLS)*

The series production version of the Abrams tank was designated the M1 in early 1981, a couple of examples being shown here. As is evident from this picture, they retained the wedge-shaped rear wall of the turret as seen on the LIRP XM1. However, the center portion of that rear turret wall, between the antenna posts, was extended 8 inches rearwards, resulting in a box-like protrusion with a vertical rear face. *(GDLS)*

Another LIRP XM1 identifying feature was the lack of a flash hider/suppressor on the coaxial (coax) 7.62mm machine gun firing port located on the left side of the main gun mantlet. The coax on the M1 tank, like the other machine guns on the tank, is employed against targets that do not warrant a main gun round. *(GDLS)*

Unlike the LIRP XM1, the series production M1 *was* fitted with flash hider/suppressor on the coaxial (coax) 7.62mm machine gun firing port. It appears in this picture as a long slender metal tube/shroud that projects out of the mantlet. At night it would prevent the muzzle flash from the coax washing out the gunner's primary sight (GPS). *(GDLS)*

(*Above*) Visible is the muzzle flash from the 105mm main gun on an M1 tank being fired. The aiming point for the tank's gunner appears as a dot within a circle in the optical image in the Gunner's Primary Sight (GPS). The aiming points are referred to as reticles. On the M1 tank and subsequent versions they appear in the gunner's optical path by way of a small projector. (*GDLS*)

(*Opposite above*) Taken at the former US Army Armor School, located at Fort Knox, Kentucky, is this picture of an M1 tank. The camouflage paint scheme visible on the tank was one of the numerous variations developed by the US Army Mobility Equipment Research and Developmental Command (MERDC) in the early 1970s. Its use was discontinued in the late 1980s. (*TACOM*)

(*Opposite below*) Depending on the terrain and environmental conditions in which they are operating, tank tracks can have a very short service life. Modern tank tracks made of steel and rubber, like those fitted to the M1 tank series, will not last more than 2,000 miles. To prolong the service life of tank tracks most armies employ large-wheeled tank transporters as seen here carrying a US Army M1 tank. (*GDLS*)

The faster way to transport the Abrams series of tanks around the world is by cargo plane. Pictured is an XM1 tank being loaded into the cargo fuselage of a C-5 series Galaxy. During peacetime the aircraft is limited to only a single Abrams series tank, for safety's sake. In wartime, it is approved to carry two Abrams series tanks. (DOD)

Chapter Two

M1 through M1A1 Abrams Tanks

Chrysler and then GDLS would build 2,374 units of the first model of the M1 tank between 1981 and 1985 for the US Army. It was sometimes referred to as the 'basic M1'. It weighed approximately 61 tons combat loaded, and on level roads could reach a maximum speed of 45 mph. Its main gun could easily engage and destroy enemy tanks at a range of a mile, day or night, stationary or on the move.

By February 1985, a month after M1 tank production ended nineteen battalion sets of fifty-four M1 tanks each had been fielded, with eleven of those being deployed to West Germany as part of the American military contribution to the North Atlantic Treaty Organization (NATO). The first M1 tank battalion set appeared in West Germany in 1982, the same year that the first battalion set of M1 tanks was fielded in the United States.

All the M1 tanks were retired from regular US Army service by 1996. Some would remain in US Army National Guard armored units until 2007. All were eventually placed into storage for possible rebuilding into newer versions if the need arose, or for conversion into specialized vehicles, such as armored vehicle bridge-launchers, armored recovery vehicles, or armored engineer vehicles.

Improving the Breed

Even as the first M1 tanks were rolling off the assembly line, the US Army was looking down the line for an improved model of the Abrams tank series. This desire was formalized in September 1981, when the Vice Chief of Staff of the US Army approved two separate developmental efforts on upgrading the Abrams tank series.

One of these developmental efforts involved the previously agreed upon integration of a German-designed 120mm tank gun into the new American tank. The US Army felt that the German-designed 120mm tank gun was a fine gun, but needed some tweaking for American use. This resulted in the production of a modified version of the German tank gun for the M1A1 designated the M256.

Despite the now combat-proven merits of the 120mm main gun on the M1A1 and M1A2 series tanks, not everybody in the US Army was originally in favor of mounting it in the Abrams tank. In a 1989 interview conducted by the US Army Center for Military History, Major General Sunell commented on the controversy surrounding the selection of the German-designed 120mm main gun for the Abrams tank:

> We knew that to go to a 120mm would cost the program a billion dollars. It has cost us that. We didn't believe that we needed the 120mm as much then as we do now. There are still those that think that we could have put the same amount of money into improving the 105mm round ... Be that as it may, the 120mm was a gun that we needed to take on the Soviets, and we have it in production. It was the right decision at the time although it cost us quite a bit of money.

Retired US Army officer James Warford comments on the need for the 120mm main gun to be fitted to the Abrams tank series:

> I've put a lot of time into studying just how good (or bad) our 105mm ammo was back in the day. The bottom line is that we would have been in a big trouble had the balloon gone up [The Third World War]. Problems, ranged from poor penetration capability, to significant round dispersion problems, with our whole family of 105mm APFSDS [Armor-Piercing, Fin-Stabilized, Discarding Sabot] rounds. If I had gone to war as a platoon leader or company commander, my 105mm ammo would have simply failed me and my men. It would have been World War II all over again. Any proposed 'new' 105mm round (after the M833) would have truly had to be a very impressive silver bullet to make a difference. We needed the 120mm as quickly as possible and most of us in the field or at the 'tip of the spear' if you prefer, didn't know it.

The second developmental effort for a second generation Abrams tank was a product improvement program (PIP), which included better armor, an improved suspension system, upgraded transmission, and a Nuclear Biological or Chemical (NBC) protection system.

Until the successful integration of these two developmental efforts into a next version of the Abrams tank series, the M1 tank modified prototypes being tested were designated as the M1E1 tank. GDLS delivered the first units of the M1E1 tanks for testing by the US Army in March 1982.

After an extensive testing of the prototype M1E1 tanks, the US Army standardized the vehicle in August 1984 as the M1A1 Abrams. The first production units of the M1A1 tank left the factory floor in 1985 and the last in 1993. The US Army currently lists the M1A1 tank as weighing approximately 68 tons. The M1A1 tank model was the last new-built Abrams series tank for the US Army. Some US Army Abrams tankers would refer to the M1A1 tank as the 'A1' if asked what specific tank they were serving on.

Upgrading the M1 with M1A1 Features

Some of the upgrades that went onto what became the M1A1 tank, minus the 120mm main gun, and the NBC protection system, were deemed suitable for fitting to the first model M1 tank. So, at the same time the M1A1 tank was being standardized, a program began to apply many of its features to the M1 tank production line. This resulted in the building of a vehicle referred to as the IPM1. The two letters in the designation code stood for 'improved.'

GDLS would build 894 units of the IPM1 between 1984 and 1986. Unlike the M1, the IPM1 was built with the ability of being up-gunned to mount a 120mm main gun. There were sixty-four differences between the IPM1 and the M1A1. The IPM1, like the original M1, is no longer in service and reserved for rebuilding into newer models of the Abrams tank series, or specialized variants.

Armor Upgrade for the M1A1

Reflecting the threat of ever-improving Soviet antitank munitions, the US Army decided to upgrade the protection on the M1A1 tank during its production run. Beginning in May 1988, all the M1A1 tanks coming off the factory floor had a layer of depleted uranium (DU) armor inserted into their existing turret armor array. A total of 2,329 units of the M1A1 tank had been built without the DU armor.

DU is a uranium byproduct that has undergone a process to remove any of the material useful in nuclear weapons or in nuclear reactor fuel. Since DU armor is more than twice as dense as conventional steel tank armor, it provided the M1A1 tanks so equipped with a greatly increased level of protection from both armor-piercing (AP) projectiles, and shaped charged warheads, referred to as high-explosive antitank (HEAT) rounds, when fired from tank guns.

The M1A1 tanks equipped with this first-generation DU armor were labeled the M1A1 HA, with the letter designation 'HA' standing for heavy armor. There are no external differences between a non DU equipped M1A1 tank and those that received the first generation DU armor. In total, 2,140 units of the M1A1 tank had the DU armor added on the GDLS production line.

The process of incorporating special armor into the M1A1 HA was not without problems, as is noted in an essay by Major General Sunell, 'The Abrams Tank System', in the book *Camp Colt to Desert Storm: The History of US Armored Force*:

> The transition from R&D [research and development] to the production of DU armor was far more difficult than anticipated. While it was relatively easy to make a few armor packages by hand, it was a much bigger challenge to make quality packages at a high production rate as DU armor had never been fabricated in that fashion before …

There was a later production version of the M1A1 HA tank that had an improved second-generation DU armor protection array fitted. That vehicle was labeled the

M1A1 HA + [plus] tank. The breakdown between the number of M1A1 HA tanks built, with the first generation DU armor, and the number of M1A1 HA + tanks built, with the second generation DU armor, is unknown.

A US Army Center for Military History publication dealing with the first war between the United States and Iraq in 1991 titled *The Whirlwind War, The US Army in Operation Desert Shield and Desert Storm* describes the armor arrangement inside the tank as '… ceramic blocks set in resin between layers of conventional armor.'

At least one television program on the Abrams tank suggests that the ceramic blocks inside the special armor portions of the vehicle function by deforming upon impact when struck thereby absorbing and dissipating the energy of an antitank projectile. Articles in some defense magazines imply that when struck by an antitank projectile the shattered ceramic blocks in the path of the antitank projectile create a cloud of small hard particles that disrupt and erode it before it can reach the rear of the armored box, containing the special armor array, which is referred to as the 'backing plate.'

A Unique M1A1 Design Feature

At some point after November 1990, a new design feature was added to the M1A1 HA tank production line. It consisted of six small adjuster wheels, three mounted on each of the tank's two roof-mounted rear turret bustle blow-off panels. They were intended to be lifting points for an overhead crane that could then remove the entire main gun ammunition rack when empty from the rear turret bustle and then quickly insert another full main gun ammunition rack, with 40 rounds, into that space.

For whatever reason, the quick reload plan envisioned by the USMC for their fleet of M1A1 HA tanks built after November 1990 was cancelled; the date this occurred is unknown. It is also not known if this was a specific USMC project, or if it was ever considered by the US Army. Despite the cancellation of this program the six adjuster wheels seem to have remained in production and show up on both USMC M1A1 tanks, as well as US Army M1A1 HA and M1A1 AIM tanks.

The adjuster wheels do not appear on the newest versions of the M1A1 series tanks. Neither do they show up on the M1A2 series tanks, which replaced the M1A1 series tanks in US Army use. However, a close-up examination of the blow-off panels on these newer generation tanks shows the six circular outlines for the cutting out and mounting of the adjuster wheels.

Into Combat with the M1A1 HA Tank

By the time the ground combat operation phase of Operation Desert Storm started on February 24, 1991, the US Army had swapped out all the M1 and most of its IPM1 tanks sent to the theater in the previous months with the M1A1 tank. Those in theater therefore consisted of a brigade of IPM1 tanks and the more numerous M1A1

HA tanks as well as the pre-May 1988 built units that had not been fitted with the DU armor.

To ensure that all the M1A1 tanks in the Middle East would have the same degree of protection, a crash program was instigated prior to ground combat operations, which brought over 800 units of the early production units of the M1A1 up to the M1A1 HA standard. Eventually, the majority of the 2,388 pre-May 1988 built units of the M1A1 tanks that had not originally been fitted with DU armor at the factory were brought up to the M1A1 HA or M1A1 HA + standard.

During a number of combat encounters between US Army M1A1 HA tanks and the Soviet-designed tanks that equipped the Iraqi Army, the American tanks were fired upon with armor-piercing (AP) main gun rounds by Iraqi tanks at a variety of ranges. None of the rounds that had impacted on the frontal armor array of the M1A1 HA tanks managed to penetrate. In fact, even when struck by friendly fire from other M1A1 HA tanks, the frontal armor array of the M1A1 HA tank could not be penetrated.

Besides its superior armor protection, another crucial advantage enjoyed by the M1A1 HA tank over its Iraqi opponents during Operation Desert Storm was its more sophisticated fire-control system. This had been the number two desired characteristic identified by the MBTTF after protection, which had been listed as the number one characteristic sought for the US Army's future MBT.

The superior fire-control system on the M1A1 HA tank allowed its crews to engage and destroy Iraqi tanks long before they were within the effective range of their own tanks' much cruder fire control systems. The fact that American tank crews were much better trained than their Iraqi counterparts also did much to produce the lop-sided results in the very short 100 hours of the ground campaign phase of Operation Desert Storm, with no M1A1 HA tanks being lost to enemy tank fire, and a countless number of Iraqi tanks destroyed by M1A1 HA tank fire.

In a book titled *War Stories of the Tankers*, published in 2008 and edited by your author, comes this passage by US Army Colonel (now Lieutenant General) H.R. McMasters. It contains his explanation for the overwhelming advantage the M1A1 HA enjoyed over the Iraqi Army's best tank, the T-72, during the famous Battle of 73 Easting that took place during Operation Desert Storm. His unit led the attack which overwhelmed the Iraqi armored units:

> Overmatch on the ground also stemmed, in large measure, from technological superiority. The M1A1 tank completely outclassed the T-72. It is impossible for a well-trained M1A1 crew to miss with service ammunition. The round goes right in the aiming dot at the most extended ranges. Everything hit is catastrophically destroyed... Our stabilized weapons systems that permit us to fire on the move with absolute accuracy allowed us to close with the enemy and achieve shock

effect as we destroyed large numbers of enemy vehicles and positions. Our optics and especially our thermal sights gave us an advantage in acquiring the enemy, especially in limited visibility conditions. Finally, the global positioning system (GPS) permitted us to navigate in featureless desert under tough conditions, including sandstorms.

Desert Storm Abrams Shortcomings

All those who served on the Abrams tanks during Operation Desert Storm were greatly impressed by the power and performance provided by the vehicle's gas turbine engine. However, this came at a cost as the gas turbine engine on the Abrams tends to use more fuel compared to past and current diesel tank engines. This issue was addressed in a United States General Accounting Office report issued in January 1992 titled *Operation Desert Storm; Early Performance Assessment of Bradley and Abrams*:

> High fuel consumption limited the tank's range, and refueling the tank was a constant consideration in operational planning throughout the ground war. Tanks were refueled at every opportunity in order to keep the fuel tanks as full as possible … Although these efforts provided optimum fuel availability, almost everybody we interviewed agreed that the tank's high fuel consumption was a concern. Typically, those we interviewed said that high fuel consumption was a trade-off for increased power and speed but that fuel economy could be improved by the addition of an auxiliary power unit.

Auxiliary Power Unit

The US Army had long anticipated the need for an external auxiliary powered unit (EAPU) for the Abrams tank and had wanted to install it on the M1 tank but it could not be included at the time due to cost. The search for a suitable EAPU began again in 1983.

The first version of an EAPU for the Abrams tank was a diesel-powered unit positioned in an enclosed box on the right rear of the tank's hull. Based on pictorial evidence a number of these were mounted on Abrams series tanks that saw service during Operation Desert Storm. Crew response to the EAPU was not positive and their location at the rear of the tank hull proved less than optimum. They were not seen after Operation Desert Storm. However, the two mounting brackets that supported this first EAPU appeared on some of the M1A1 HA tanks that took part in Operation Iraqi Freedom.

Based on the experience of Operation Desert Storm the US Army developed another type of diesel-powered EAPU that was small enough to fit within the confines of the turret bustle on M1A1 HA tanks. These are seen on the M1A1 HA tanks that took part in Operation Iraqi Freedom in 2003. They continue to be employed on the current version of the M1A1 series tank employed by the US Army National Guard and US Marine Corps.

Keeping the M1A1 HA Tank Fit for Service

Due to the hard usage of its M1A1 HA tank fleet, in both war and training exercises, the US Army decided in 1994 to begin a rebuild and upgrading program for a portion of its M1A1 HA tank inventory to prolong its service life. The rebuild program was originally assigned the name Abrams Integrated Management for the twenty-first century (AIM XXI). At some point in time, the 'XXI' was dropped from the acronym and the rebuild program was simply referred to as AIM.

The first seventeen units of the M1A1 HA tank completed under the AIM program went out for testing in 1997 to determine the effectiveness of the process. The test results were extremely positive resulting in the US Army approving the program to go forward in June 1998.

By 2012, the AIM program was completed, and approximately 2,000 units of the M1A1 HA tank had cycled through it, becoming known as M1A1 AIM, to distinguish them from M1A1 HA tanks that had not gone through the program.

Information Based Systems and the Abrams Tank

Among the features originally added to some of the M1A1 AIM tanks were the Force XXI Battle Command and Brigade and Below (FBCB2) and Blue Force Tracking (BFT). Initially these would have been intended for command tanks, but eventually fitted on all M1A1 AIM tanks.

In the simplest of explanations, FBCB2 is a wireless tactical internet system. It takes real-time information, which it receives from different sources, and transmits it over digital radios. The end result is that tank commanders achieve an unheard-of degree of situational awareness by looking at a touch-screen color monitor within their tanks that displays a moving digital color map with tactical symbols. No longer do they need to deal with paper maps and compass.

First Sergeant Antoine Haddar, of the Army's 4th Infantry Division explains what the FBCB2 brings to the battlefield in his opinion:

> It gives us situational awareness. Before, you had to listen to the radio carefully, and you couldn't get all the information. Now, you can basically see everything, [and] the whole brigade can see the same thing. All that information used to be privileged, and you got surprised. Now, you get surprised less, actually a lot less, and this is an advantage.

From the 3rd Infantry Division (Mechanized) After-Action Report on Operation Iraqi Freedom, posted online, appears this quote: 'The FBCB2 "magic box" was one of the most valuable items that we used during the war. It creates a common understanding instantly and shows you in relation to your brother. Many times I was able to visualize a FRAGO [Fragmentary Order] as it was being given.'

BFT is a Satellite Communication (SATCOM) device, the antenna for which is located on a platform that projects out of the right side of the M1A1 AIM tank turret. It provides the tank crew the location of all friendly ground and aerial platforms. According to an online document from the US Army Abrams Tank Program Manager's office, titled *Lessons Learned Operation Iraqi Freedom 2003*, appears this extract: 'Satellite based communications proved to be the most reliable form of communication.'

The application of FBCB2 and BFT to the US Army's tank fleet reflected the growing importance of real-time information to the army of the twenty-first century. This is seen as a US Army concern as far back as 1994 when the then Chief of Staff of the US Army, General Gordon R. Sullivan stated the following:

> The high ground is information. Today we organize the division around the killing system, feeding the guns. Force XXI must be organized around information ... The creation and sharing of knowledge followed by unified action based on that knowledge which will allow commanders to apply power effectively ...

Updating the Abrams Engine

There have been no new AT-1500 gas turbine engines built for the Abrams tank series since 1994. Plans had been floated at different times to build a brand new gas turbine engine, or even to install a diesel engine in the tank as favored by GDLS. None of these plans ever came to fruition. Instead, the US Army decided in 2006 that the most affordable and cost-effective approach to keep the existing AT-1500 gas turbines in service was to upgrade them with new parts to extend their service life, bringing the engine to a 'zero miles/zero hours' configuration. This engine upgrade program was assigned to the firm of Honeywell and called the Total Integration Engine Revitalization, or 'TIGER.' It was completed in 2014.

Operation Iraqi Freedom

When the time came for Operation Iraqi Freedom in 2003, the mainstay of the ground element of the US Army that invaded Iraq consisted of approximately 450 M1A1 tanks as well as a newer model labeled the M1A2 tank. It is unclear how many of the M1A1 tanks that took part in the fighting in Iraq could be classified as either M1A1 HA or M1A1 AIM tanks, so I will use the term M1A1 for the vehicles which fought in Iraq at that time.

As they did during the ground phase of Operation Desert Storm, the American Abrams tank series completely dominated their Iraqi tank counterparts in every combat engagement, until the Iraqi government and military collapsed within the span of three weeks.

An example of the effectiveness of the M1A1 series tanks in combat during the first couple weeks of Operation Iraqi Freedom appears in this extract from the official US Army history of that conflict titled *On Point, US Army in Operation Iraqi Freedom*:

> As soon as Waldron sent his short contact report to Charlton, he issued a platoon fire command to his lead platoon. With his tank adding its firepower to the four others in the platoon, in less than 30 seconds after the radio call, the massive 120mm cannons on five tanks roared in unison. The firing continued for two minutes as the gunners and tank commanders traversed left and right, seeking out and destroying the tanks and other vehicles dug into supporting positions. In less time than it takes to tell, they destroyed four T-62 tanks, several other armored vehicles, and some trucks that were moving behind some bunkers.

From the 3rd Infantry Division (Mechanized) After-Action Report on Operation Iraqi Freedom, which can be found online, appears this summary on the effectiveness of the division's armored fighting vehicles, especially the Abrams series tanks:

> This war was in large measure won because the enemy could not achieve effects against our armored fighting vehicles. While many contributing factors, such as air interdiction (AI), close air support (CAS), Army aviation, and artillery helped shape the division battle-space, ultimately any war demands closure with an enemy force within the minimum safe distance of supporting CAS and artillery. US armored combat systems enabled the division to close with, and destroy heavily armored and fanatically determined enemy forces with impunity, often within urban terrain. Further, the bold use of armor and mechanized forces striking the heart of the regime's defenses enabled the division to maintain the initiative and capitalize on its rapid success in route to Baghdad. During MOUT [military operations urban terrain], no other ground combat system currently in our arsenal could have delivered similar mission success without accepting enormous casualties.

The director of the RAND Corporation's Center for Middle East Policy said: 'It turns out that the tank we thought we were going to fight the Russians with is the best thing we've got to fight in an urban environment.'

Despite the brilliant showing of the Abrams series tanks during the three week campaign that ended Saddam Hussein's rule of Iraq, there were some shortfalls in gunnery by American tank crews. This passage from an article titled '1-64 Armor Rogue Training Program' published in the January–February 2004 issue of *Armor* magazine addresses those issues:

Any analysis of potential future combat scenarios leads to the conclusion that enemy forces will seek to fight in close terrain ... The armor community can no longer afford to focus primarily on the conventional fight. We must adapt our training to likely future scenarios and ... tank gunnery must adapt to the changing threat environment.

The Iraqi Insurgency

Following the successful conquest of Iraq by the American military forces in 2003, fighting continued in the country. It was the US Army Abrams tank series which provided the key asset in trying to suppress the violent urban warfare environment which engulfed the nation. In a book titled *War Stories of the Tankers*, published in 2008, and edited by your author, comes this passage by then US Army Captain (now Lieutenant Colonel) John C. Moore, describing what it is like being struck by an RPG-7 warhead in your M1A1 series tank:

> A parachute flare went up to our south and drifted noiselessly downward. We clearly had not fired it. I had my head just out of the hatch enough to see over the vision blocks when an RPG slammed into the right side of my tank. The blast pushed me down in the hatch, and the driver, Spec. Peter Zuhan, screamed out that we had been hit. I am sure the concussion was severe in the sealed driver's compartment. Other gunfire commenced, and the tanks responded with every weapon available. RPGs zipped in from the west and the east. My tank was hit again ... I was concerned that we were a mobility kill, unable to move because of damage to the suspension or track ... All four tanks were damaged but all of them were still mission capable. The M1A1 is an absolutely amazing piece of equipment.

From a US Army Center of Military History publication titled *Tip of the Spear: US Army Small-Unit Actions in Iraq, 2004–2007* appears this extract showing the intensity of the fighting during the Iraqi Insurgency:

> By 0500, four Company B tanks, commanded by Lieutenant Palani and Sergeants Franklin, Wright, and York, were deployed along a screen line south of Route Huskies. The insurgents, intent on reoccupying ambush positions along the main highway, launched a coordinated RPG attack at 1000 against the tanks. An RPG team succeeded in hitting a road wheel on the right side of York's Abrams. None of the other vehicles were struck. All of the tanks began searching for their tormentors. Anxious to avoid being hit a second time, Sergeant York shifted his tank several hundred yards to the west. Moments later, he observed seven enemy armed with RPGs and AK47s. York wiped out the group using his coaxial machine gun and turret-mounted .50-caliber. York encountered

another seven-man RPG team while he repositioned his vehicle yet again. Although both sides were surprised, the Iraqis got off the first shot, hitting the M1A2 with an RPG round that pierced the armored skirt protecting the suspension and track. The sergeant responded seconds later with a single main-gun round that killed everyone in the enemy force.

In the midst of yet another shift in location, York saw a third RPG team moving across an open area. The insurgents likewise spotted the tank and attempted to hide behind a building. Sergeant York fired a main-gun round into the structure, killing the enemy. Three men began shooting at York as he leaned out of his commander's hatch in an effort to pinpoint a safe route for his vehicle. Dropping back down into the turret, York engaged each in turn, killing two. Eight insurgents, including two RPG gunners, fired at the tank from a nearby alley. York responded with coaxial and .50-caliber machine-gun fire, wiping them out before they could reload their launchers.

York's fire and maneuver stirred up a hornets' nest. Sergeant Franklin saw a large number of enemy dash behind several houses. Unable to bring his own weapons to bear, Franklin pointed out their location to York, who used an M203 grenade launcher to force them from cover. The tactic succeeded as Lieutenant Palani, Sergeant Franklin, and Sergeant Wright observed insurgents fleeing to the east. A fusillade from all three tanks killed or wounded twenty.

According to a 18 February 2005 article in *Army Times*, US Army Major General Terry Tucker stated that over 1,100 Abrams tanks had seen action in Iraq, and of that number, 70 percent had been struck and damaged in some manner. Tucker also stated that 80 Abrams tanks had suffered enough damage in Iraq to warrant the process of shipping them back to the United States. Of that number, 63 were to be repaired and placed back in service, with the remaining 17 tanks not worth fixing.

Major General Tucker went on to explain in the 2005 article the tactics the Iraqi insurgent employed when engaging Abrams tanks: '... the technique that they're using is massed fire against one tank: 14, 18, 20 RPGs – I've heard reports of 50 RPGs hits. It's a new technique that they're using, and in fact we're having some significant damage on tanks that had to be repaired before we put them back in the fight.'

Besides RPGs, the Iraqi insurgents employed Improvised Explosive Devices (IEDs). At one point, the insurgents improvised an antitank mine consisting of a 55-gallon oil drum packed with C-4 plastic explosives. An Abrams tank triggered the mine and the resulting blast killed three members of the tank's crew and the vehicle lost its turret.

The M1A1 SA Tank

Beginning in 2009, the M1A1 AIM tanks in the US Army National Guard inventory underwent some additional improvements. By this time, the US Army was only using the M1A2 tank series. These additional improvements included upgrades to the side armor on the tank's turret, and a phone, mounted within an armored box, added to the right rear of the hull of the vehicle. The phone's official designation is 'TIP,' which stands for Tank-Infantry Phone.

Reflecting the upgrades applied to the monitoring and sensor technology of the M1A1 AIM tanks the US Army relabeled them the M1A1 SA tank, the letters 'SA' standing for situational awareness. Some references have appeared that label them as the M1A1 SA AIM tanks.

Beginning in 2012, the US Army National Guard's inventory of M1A1 SA tanks were fitted with a new feature known as the Stabilized Commander's Weapon Station (SCWS). It no doubt reflected the lessons learned from the urban combat experience gained by the American military in Iraq, where the lack of a stabilized CWS on the Abrams tank series was seen as a disadvantage when engaging enemy ground personnel. A stabilized CWS had been desired on the M1 tank but not added due to cost.

The SCWS is stabilized, like the vehicle's turret, in both azimuth and elevation, and can be accurately fired by the tank commander from within his vehicle, or when he decides to project his head and upper torso out of the turret for better visibility. Additional protection is afforded the tank commander by transparent armor panels on either side of his position when exposed. Rear protection for the tank commander, when his head and upper torso is exposed, is provided by his overhead hatch set in a vertical position.

Six US Army National Guard Armor Brigade Combat Teams (ABCTs), formerly referred to as Heavy Brigade Combat Teams (HBCT), are equipped with the M1A1 SA tank. Each ABCT has fifty-five tanks. Also equipped with the M1A1 SA tanks are three separate US Army National Guard Combined Arms Battalions.

(*Opposite above*) The original rear hull skirt plate design is seen here on a series production M1 tank. It covered half of the rear drive sprocket. Sometimes it caused mud to build up at the two drive sprocket positions, which resulted in some early production M1 tanks throwing their tracks. The rapid backing up and turning of the early production M1 tanks also caused tracks to be thrown. (*GDLS*)

(*Opposite below*) To resolve the problem of mud building up under the original rear hull skirt plates of M1 tanks, which often causing tracks to be thrown, a number of fixes were implemented. The easier option for many tank crews was the simple expedient of removing the rear hull skirt plates on their vehicles as seen here. Another more involved design fix was the addition of a 25-inch track retainer plate over the rear drive sprocket as shown in this picture. (*Michael Green*)

(*Above*) An M1 tank is shown descending a muddy hill, which has always been a tricky proposition for the novice driver. In training, Abrams series tankers are taught to drive aggressively and to learn how to exploit battlefield situations and turn them to their tank's advantage. When engaging enemy tanks they are instructed to seek out hull-down positions that protect their vehicle from enemy fire but at the same time to allow their tank's gunner to put maximum fire on the target. (*GDLS*)

(*Opposite above*) Taking part in training maneuvers, as is indicated by the flash simulator over the main gun barrel, popularly referred to as the 'Hoffman device,' are several M1 tanks. The vehicle was 32 feet long, 7 feet 9 inches tall, and had a width of 12 feet. Maximum weight of the tank was 60 tons. (*DOD*)

(*Opposite below*) The crew of the M1 tank and all subsequent models consists of four men, as seen in this ghosted diagram. The driver sat in the front hull and the other three in the turret: the tank commander and gunner on the right hand side of the vehicle, when looking forward from the rear of the tank, and the loader on the left side of the main gun. (*DOD*)

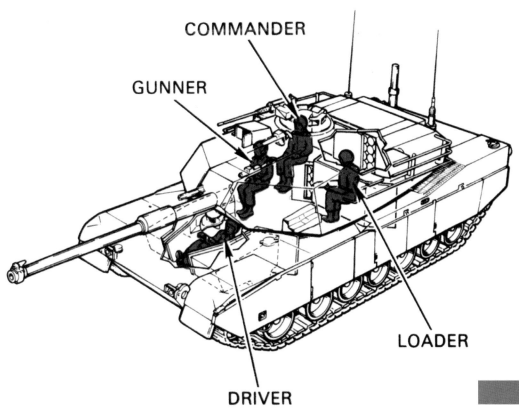

(*Above*) Two M1 tanks are pictured during a training exercise at Fort Hood, Texas. In spite of the massive angular appearance of the turret, it had less interior space than the M60 series tank turrets. The vehicles are both painted in a four-tone MERDC camouflage scheme labeled 'Gray Desert.' Unit markings are done in black. The paint on the vehicles has infrared suppression qualities. (*GDLS*)

(*Opposite page*) The biggest shock too many in the US Army armored community with the advent of the M1 tank (a number are seen here) was how quiet its gas turbine was in operation. Even at full power, it only produced a muted whine. During early training maneuvers in Western Europe the tank was nicknamed the 'Whispering Death.' (*GDLS*)

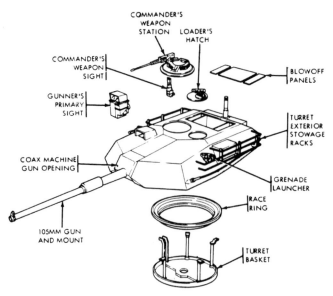

An important safety design feature on the M1 tank were the three blowout panels located on the roof of the vehicle's turret bustle, as seen in this diagram. If an enemy projectile penetrated the turret bustle and detonated the 44 main gun rounds stored within, the resulting ammunition explosion would be vented upward and outward and not into the vehicle's crew compartment. *(DOD)*

This picture taken from the gunner's position in an M1 tank shows the loader's seat. To the left of the loader's seat is one of the vehicle's two power-operated sliding armor blast doors. These doors separated the crew of the vehicle from the main gun rounds stored within the turret bustle, in case of detonation. The 22 main gun rounds stored behind the blast door pictured were labeled the 'ready rack.' *(Michael Green)*

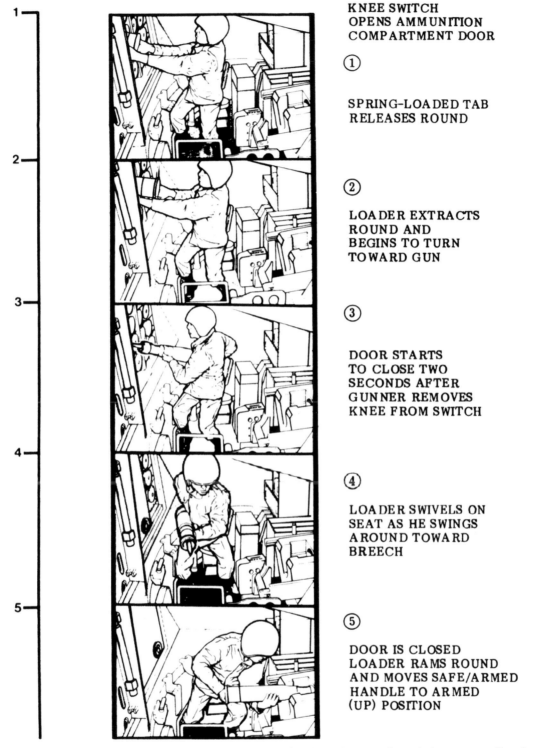

A line drawing shows the various events that transpire as an M1 tank loader goes through the process of loading the tank's main gun from the ready rack. The same process is found on all subsequent models of the Abrams tank series. Once the loader uses his knee switch to open the power-operated sliding armor blast doors, he releases a spring-loaded tab holding a round in place, extracts a round, and turns towards the breech of the main gun with it. (DOD)

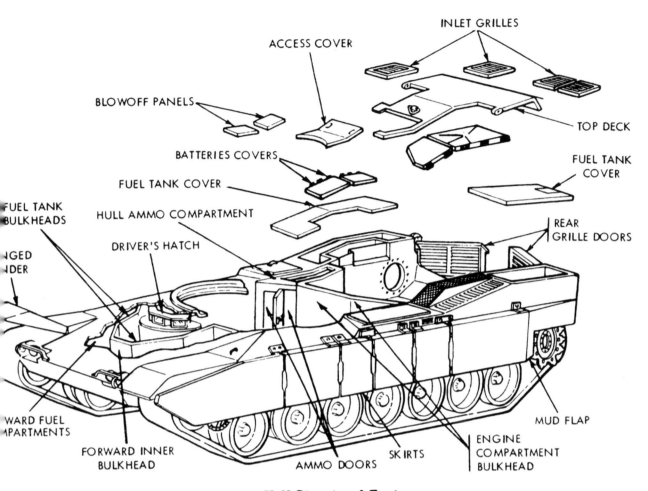

Hull Structural Features

(*Above*) Many of the various components that made up the M1 tank's hull are shown in this line illustration. Besides the 44 main gun rounds stored in the turret bustle, there were another 11 stored in the vehicle's hull; 8 of these were stored behind blast-proof doors in the hull and another 3 were located in an armored box on the turret floor. (*DOD*)

(*Opposite above*) This wider interior view of the rear of an M1 tank turret shows the tank commander position on the left hand side of the picture. Behind the tank commander's up-turned seat is a white metal shield. Behind it is a second power-operated sliding armor blast door. In that portion of the turret bustle is stored another 22 main gun rounds, referred to as the 'semi-ready rack.' (*Michael Green*)

(*Opposite below*) In this picture of the rear of an M60A1 tank turret we see that the crew of the vehicle has no protection whatsoever from the onboard main gun ammunition if detonated. The danger of this main gun ammunition storage arrangement became painfully obvious to the Israeli Army during the 1973 Yom Kippur War, when it lost many of its American-supplied M60A1 tanks to internal main gun ammunition detonation. (*Michael Green*)

Visible in this photo taken from the loader's position in an M1 tank turret are the gunner's and tank commander's positions. The Gunner's Primary Sight (GPS) is located to the right of the 105mm main gun breech. The gunner's computer control panel, with its door open in this picture, is to the right of the GPS. Just above the gunner's GPS is the tank commander's GPS extension. *(Michael Green)*

This line drawing shows the various positions and components that made up the gunner's position on an M1 tank. This arrangement has remained a near constant on subsequent models of the vehicle for the gunner, despite numerous improvements and add-ons. Despite considerable engineering to ergonomically improve the gunner's position, it remains the most cramped and uncomfortable in the tank. *(DOD)*

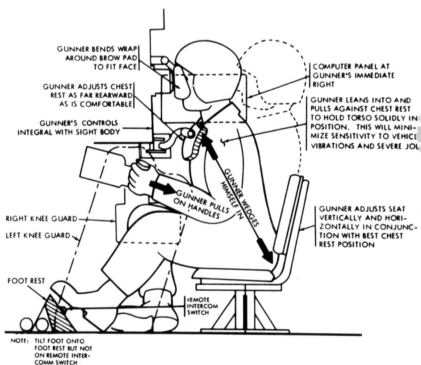

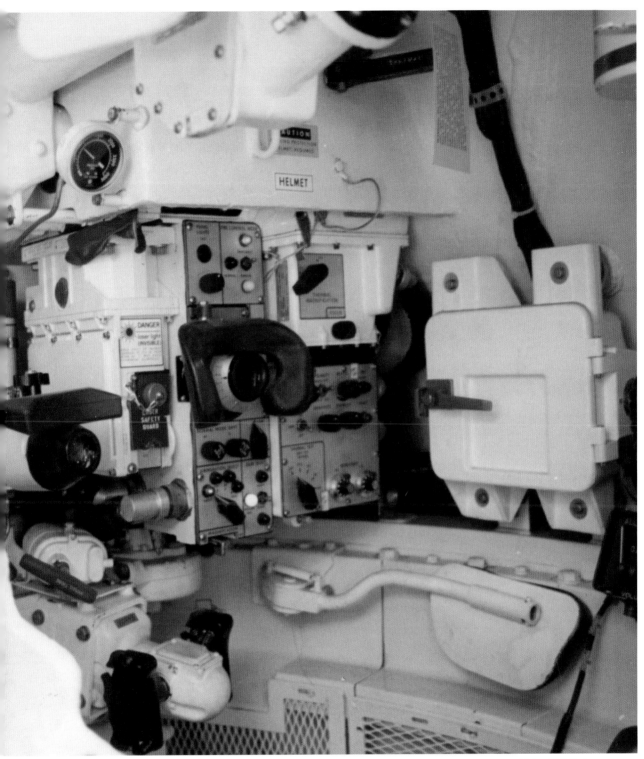

In this close-up picture of the gunner's station on an M1 tank we clearly see the Gunner's Primary Sight (GPS). Just below the GPS are the gunner's control handles, nicknamed 'Cadillacs.' To the right of the Cadillacs is the folded back adjustable chest rest for the gunner. The GPS provides the gunner both an unfiltered daylight view outside the tank and also a thermal mode as well. The thermal view for the gunner is provided by the Thermal Imaging Sight (TIS) incorporated into the GPS. (*GDLS*)

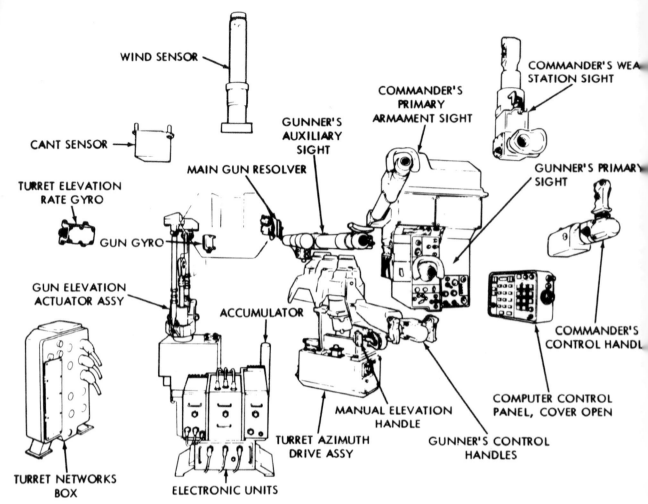

(Above) The various M1 tank fire-control system components seen in this line drawing all play a part when the gunner seeks out and engages a target. The electronic units located on the bottom left of the line drawing are located under the main gun. They provide the control signals to the azimuth and elevation servomechanisms to move the tank's turret and main gun, based on input information from other components of the vehicle's fire-control system. *(DOD)*

(Opposite above) At the moment of firing, a fireball is created by the combustible gases that hurl the projectile down the gun tube of an M1 tank. The 105mm service kinetic-energy (KE) main gun rounds used either tungsten alloy penetrators or depleted uranium (DU) penetrators. One of the latter was designated the M833 and was supposed to be able to push through 16.8 inches (420mm) of steel armor. *(GDLS)*

(Opposite below) A US Army tanker is shown carefully laying down a 105mm service kinetic-energy (KE) main gun round for an M1 tank. Colors indicate the use of a main gun round. When both the projectile, that part of the round that travels through the gun tube, and the sub-projectile, the part of the projectile that travels to the target, are black, it is an armor-defeating round. In this picture an olive drab nose cone protects the sub-projectile tip from damage. *(DOD)*

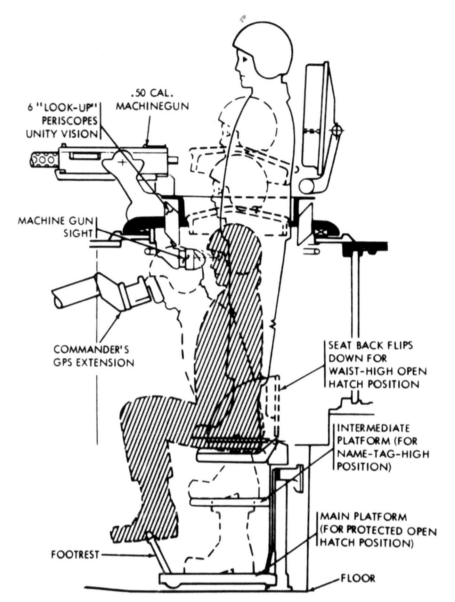

(Above) This line drawing shows all the various positions that the M1 tank commander can use at his position depending on his needs and tactical situation. As with all the crew positions on the M1 tank the vehicle's designers spent a great deal of time and effort in maximizing the available space to meet all the crew's perceived needs. To accomplish this goal, the designers questioned both active duty and retired tankers on what they felt had been lacking in past vehicles. *(DOD)*

(Opposite above) This picture taken from the loaders' position on an LIRP XM1 tank shows the tank commander, who is wearing what is referred to as a Combat Vehicle Crewman (CVC) helmet. It has a boom microphone that projects out from the right side of the helmet. Directly behind the tank commander, and parallel with his face, is the commander's daylight only weapon station sight. *(DOD)*

(Opposite below) A close-up picture of the low-profile Commander's Weapon Station (CWS) on an M1 tank, and the permanently-attached pedestal mount for a .50 caliber M2HB machine gun. The CWS can in theory be rotated 360 degrees. In practice, the CWS is limited in traverse because it can be hemmed-in by the loader's open hatch and machine gun pedestal, as well as the two antennas at the rear of the turret roof. *(Michael Green)*

An LIRP XM1 tank commander is shown standing in his Commander's Weapon Station (CWS). Just in front of his chest is the armored-protected housing for the daylight only optical gun sight he would use when operating the .50 caliber M2HB machine gun from inside the vehicle. There was a hand-held power control unit he could use to traverse the CWS when he was exposing his head and upper torso out of the CWS. (TACOM)

In this picture taken from the front turret roof of an M1 tank looking rearward is the Commander's Weapon Station (CWS) on the left. To the right of the CWS is the loader's hatch. The loader's hatch is surrounded by a skate rail. Attached to the skate rail in this picture is the pedestal for mounting an M240 7.62mm machine gun. Visible on the roof of the rear turret bustle are the three blow-off panels. *(Michael Green)*

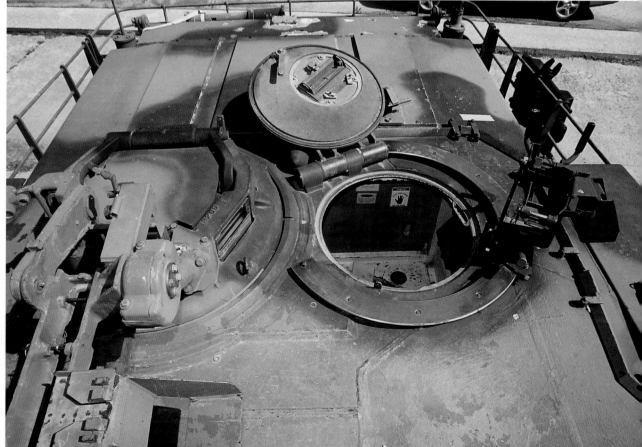

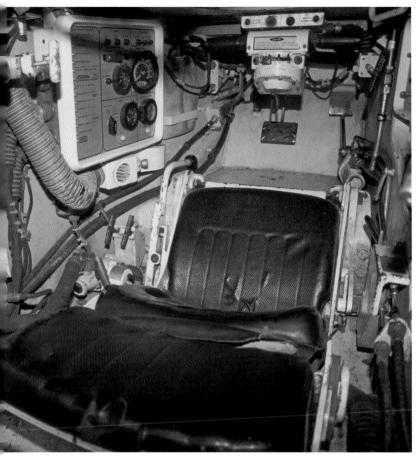

Shown is the semi-reclining driver's seat of an M1 tank. It is also found on all subsequent models of the vehicle. At the top of the picture is the T-bar steering system that incorporates motorcycle-style hand throttles. This arrangement eliminates the need for a floor accelerator pedal. As with a car, the brake pedal and parking brake pedal are operated by applying foot pressure. *(Michael Green)*

A better view of the semi-reclining driver's seat arrangement on an M1 tank, nicknamed the 'Lazy Boy' by the tankers, can be seen in this picture of a driver's training simulator. It was the British Army that first employed this driver seat arrangement on their Chieftain tank, which the US Army copied for the M1 tank, as it allowed the vehicle's height to be lowered 4 inches. *(Michael Green)*

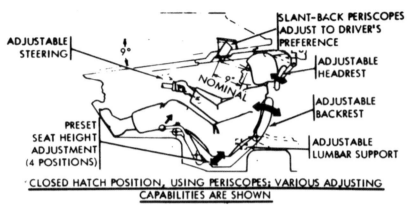

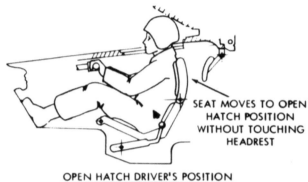

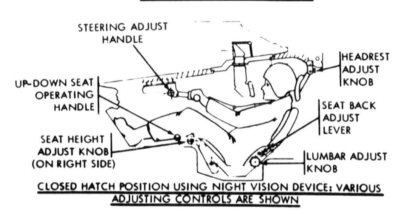

(*Above*) A line drawing shows all the various seating positions for the M1 tank driver. For night-time operations, the driver was provided with a passive night sight that would replace his center daylight-only periscope. The driver's seat on the M1 and all other subsequent models was designed to accommodate almost any size soldier, even when wearing bulky, artic cold-weather gear. (*DOD*)

(*Opposite above*) Visible in this picture of the overhead armored driver's hatch on an IPM1 tank are the driver's three periscopes, which provide a 120 degree overlapping view looking forward over the glacis plate. The color tinge seen on the center periscope is a reflection from the optical coating placed on the periscope face to prevent the driver from being blinded by laser light. (*Don Moriarty*)

(*Opposite above*) This picture shows the very large opening for the driver on all of the Abrams tanks, to ease their entry and exit from the vehicle. In sharp contrast, the first generation of M4 series medium tanks called for both the driver and the bow gunner to twist sideways before trying to squeeze in and out of their very narrow overhead hatches. (*DOD*)

(*Above*) Even before the M1 tank entered series production the US Army had decided in 1976 that it would eventually be fitted with a German-designed 120mm main gun in the name of NATO standardization. The American redesigned model of the German gun is shown here on a pilot Abrams tank designated the M1E1. In addition to the larger bore the M1E1 featured a host of other improvements. (*GDLS*)

(*Opposite above*) With testing of the M1E1 tank being considered successful, the vehicle was designated the M1A1 tank. As series production of the M1A1 tank began pushing out sufficient numbers of the vehicle they were rushed to Western Europe as replacements for the older generation M1 and IPM1 tanks, which were all brought back to the United States. By 1989, all US Army tank units in West Germany were equipped with the M1A1 tanks as seen here. (*GDLS*)

(*Opposite below*) Two M1A1 series tanks from the 1st Armored Division are seen in downtown Baghdad upon the collapse of the Iraq government in 2003. In lieu of the 120mm main gun it was armed with, there were other weapons considered for arming the M1A1 tank early on. These included the possibility of an Electro-Thermal-Chemical (ETC) gun, a liquid propellant gun, or an electro-magnetic gun. Eventually, research on these weapons was abandoned. (*Michael Green*)

(Above) Series production M1A1 tanks were fitted with a turret basket, seen here at the rear of the tank's turret bustle. This was not a feature seen on the M1E1. Eventually, the turret basket was retrofitted to the M1 and IPM1 tanks. Late production M1A1 tanks appeared with a circular projection on the turret roof, in front of the loader's position, sometime referred to as the 'manhole cover.' It was intended for the fitting of a second sighting system that did not appear until the M1A2 tank. (GDLS)

(Opposite above) The build-up of mud on the rear drive sprockets of the M1 tank due to it collecting under the rearmost hull skirt plates also plagued the early production M1A1 tanks. As with M1 tank crews, the crew of this US Army M1A1 tank solved the problem by simply removing the rearmost side hull skirt plate from their vehicle. The exposed turret basket of the vehicle pictured has been covered with a fabric tarp, nicknamed the 'burrito roll,' to protect the contents from the ravages of inclement weather. (Michael Green)

(Opposite below) In this picture of a US Army M1A1 tank being loaded onboard a US Air Force C-5M Super Galaxy transport plane, we see the final solution for the mud build-up under the rearmost side hull skirt plates of the M1A1 tank. Instead of the crude field expedient method of cutting back the non-ballistic rearmost rear side hull skirt plates, a factory-made non-ballistic version seen here was eventually applied to all M1A1 tanks. It was also backdated to the M1 and IPM1 tanks. (DOD)

(*Above*) Tankers are shown removing an AGT-1500 gas turbine engine from an M1A1 series tank. Maintenance on the AGT-1500 gas turbine engine is much easier, compared to the diesel engines on the M60 series tanks, due to its modular construction. It also has numerous quick disconnects, which allows it to be removed from its engine bay in approximately an hour, sometimes less. (*DOD*)

(*Opposite above*) A key external spotting feature of the M1A1 tank is its large bulged bore evacuator, mounted halfway down the gun barrel, as seen here on this US Army M1A1 tank in West Germany. The bore evacuator on the gun barrel of the 105mm main gun mounted in the M1 and IPM1 tanks was both smaller and rounded. All Abrams tanks have a two-piece aluminum thermal shroud as seen on this tank's gun barrel. (*DOD*)

(*Opposite below*) A tank's ability to travel off-road at high speed is dependent not only on engine power, but also the vehicle's suspension. The M1A1 tank shown here has a hydro-mechanical suspension system. It consists of roads arms that connect to high strength steel torsion bars. To prevent the vehicle from pitching when running at high speed off-road, it also has rotary hydraulic shock absorbers. The same arrangement was found on the M1 and IPM1 tanks. (*Michael Green*)

(*Above*) Mounted in the bustle rack of an M1A1 Abrams tank is a large diesel-powered External Auxiliary Power Unit (EAPU). As the gas turbine engine in the Abrams series tanks burns the same amount of fuel at high speed or idle, the APU is employed when the main engine on the tank is turned off to conserve fuel. It charges the vehicle's batteries so all of the vehicle's electrically-powered devices can continue to function. (*TACOM*)

(*Opposite above*) Seen here is the empty rear engine compartment of an M1A1 tank being license-built in Egypt. There were a number of reasons the US Army selected a gas turbine engine for the Abrams tank back in the 1970s. The US Army had been impressed by their power and reliability in helicopters. Gas turbine engines were smaller, weighed less, and it was anticipated that they would require less maintenance then existing diesel tank engines, hence would be more affordable to operate. (*GDLS*)

(*Opposite below*) During the three-week campaign that brought the Iraqi government and military to collapse during Operation Iraqi Freedom in 2003 there were no losses of Abrams tanks due to Iraqi direct or indirect fire weapons. However, several Abrams tanks, as seen, were lost to secondary effects attributed to enemy weapon systems. In some instances, improperly stored fuel for the APU dripped down into the engine compartment and caught fire, setting the engine on fire and causing the subsequent loss of the tank. (*DOD*)

Never anticipated when designed, nor taught to generations of armor crewmen who served on the Abrams series tank, was its role as a stationary pillbox to guard checkpoints in Iraq as seen here, or guard truck convoys, but it did the job. Visible is the Commander's Weapon Station (CWS) on a US Army M1A1 HA tank. The .50 caliber machine gun M2HB fitted to the CWS has a maximum effective range of 1,800 yards. (DOD)

An M1A1 tank commander is shown looking through the three-power magnification commander's weapon sight employed when he wished to fire the .50 caliber M2HB machine gun fitted to his Commander's Weapon Station (CWS) from inside the vehicle. His right hand is holding the commander's weapon station power control handle. In his left hand is the trigger for firing the weapon. (DOD)

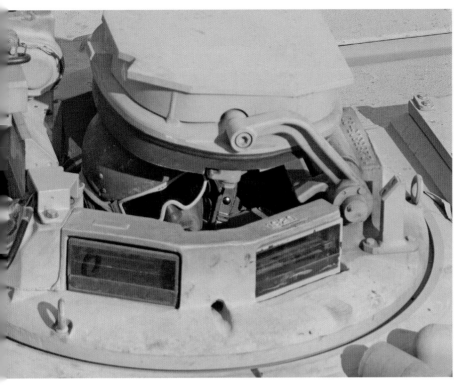

When tank commanders on the M1 through M1A1 tanks became worried about threats from above (such as small-arms fire or artillery air bursts) they could raise their overhead armored hatch to the horizontal, open protected position as seen here, which projected up just a few inches above the Commander's Weapon Station (CWS). *(DOD)*

The M240 7.62mm machine gun seen here on a US Army M1A1 HA tank in Iraq has an official maximum range of 985 yards. Beyond that range, the tracer component in the bullet burns out. Often, the firer can observe the impact of the bullets at more distant ranges and direct his fire onto the intended target. *(DOD)*

(*Opposite above*) A US Army M1A1 HA on patrol in Iraq. The headlight guards on the M1A1 tank differed from those seen on the M1 and IPM1 tanks, as they are flatter and slanted slightly forward, whereas those on the earlier models of the tank were squared off. The M1A1 tank headlight guards also have a slight dip on the top rung and have only a single attachment point, in contrast to the two attachment points seen on the M1 and M1IP tanks. (*DOD*)

(*Opposite below*) This picture of a US Army M1A1 AIM tank in Iraq shows a red primer-colored, open-topped device attached to the left hand side of the rear turret basket. It is referred to as a 'heat shield' and is employed when towing another tank. There are four upside down brackets mounted on it so just sits on the center engine exhaust grill. Its purpose is to deflect upwards the extremely hot engine exhaust gases so they do not damage the tank being towed. (*USABOT*)

(*Above*) In this close-up photograph of a US Marine Corps M1A1 series tank we can see the heat shield attached to the engine's center exhaust grill. The bracket with the holes surrounding the heat shield is for the attachment of the bottom portion of the rear engine exhaust fording tube on Marine Abrams tanks, but also appears on all US Army M1A1 HA tanks built sometime after November 1990. (*Don Moriarty*)

(Above) During Operation Desert Storm in 1991, manually cleaning the engine filters on M1A1 HA tanks as seen here was a constant preventative chore. To resolve this issue the US Army adopted a pulse jet filter-cleaning system for the M1A1 HA tanks. It works by systematically isolating each of the vehicle's three air filters and pulsing reverse air through them. This has virtually eliminated the need for manual cleaning of the engine filters. *(DOD)*

(Opposite page) Taken in Iraq is this picture of an M1A1 series tank turret that has just been lifted from its hull and is being lowered onto a maintenance stand. Visible below the turret is the tank's turret basket. The floor of the turret basket is made from aluminum and the five posts that attach to the bottom of the turret are made of steel. The tank commander's seat is attached to one of the five steel posts, as is the loader's seat. *(DOD)*

(*Above*) Tanks have a finite service life. US Army programs to develop a replacement for the Abrams series tank have foundered over the years on unresolved design issues, and ever re-sized costs. A decision was made to rebuild and upgrade a portion of its M1A1 tank inventory, as shown here. That program was labeled the Abrams Integrated Management (AIM) and took place between 1998 and 2012. (*DOD*)

(*Opposite above*) At some point in the M1A1 HA production line after November 1990 the manufacturer began fitting circular adjusting wheels, as seen in this picture, onto the two blow-off panels (three each) of US Army and US Marine Corps M1A1 series tanks. They were to be lifting points for a main gun ammunition reload project that involved the use of a crane, which was later cancelled. (*Don Moriarty*)

(*Opposite below*) The most modern version of the M1A1 tank series in service with the US Army National Guard is referred to as the M1A1 Situational Awareness (SA). Its main external identifying feature, which is not seen on earlier models of the M1A1 tanks in US Army or National Guard service, is the Stabilized Commander's Weapon Station (SCWS) shown here along with a gun shield kit for the loader's roof-mounted M240 7.62mm machine gun. (*DOD*)

(*Above*) Taken during a US Army National Guard training exercise is this picture of an M1A1 SA tank, in the foreground. In the background is an M88A2 Armored Recovery Vehicle (ARV), which entered service in 1997. At one point in time, GDLS was offering the US Army an ARV based on the chassis of the M1 Abrams tank, but the army went for the less capable but more affordable M88A2 ARV. (*DOD*)

(*Opposite above*) The rear of the Stabilized Commander's Weapon Station (SCWS) on a US Army National Guard M1A1 SA tank is seen in this picture. Just below the barrel of the .50 caliber M2HB machine gun is the black-topped plastic cover over the vehicle's Blue Force Tracking (BFT) unit attached to the side of the tank's turret. The M1A1 SA does not have the adjusting wheels seen on some of the M1A1 HA tanks, or on the M1A1 AIM tanks. (*DOD*)

(*Opposite below*) Visible in this picture is the very elaborate arrangement of chains employed to tie down a US Army National Guard M1A1 SA tank within a US Air Force C-17 transport plane. The engine exhaust grill on this tank lacks the bracket with the holes for the attachment of the bottom portion of the rear engine exhaust fording tube employed by Marine Abrams tanks. (*DOD*)

Chapter Three

M1A2 Abrams Tank

The next step in the progression of the Abrams tank series occurred in December 1988 when GDLS received the contract for full-scale development of an upgraded Abrams series tank to be designated the M1A2 tank. The firm delivered the first five prototypes to the US Army in March 1992. Impressed by the test results of the prototype, the US Army ordered sixty-two series production units of the M1A2 tank for additional testing. These vehicles were supplied to the service between November 1992 and April 1993.

The US Army would have preferred a new tank in the early 1990s, rather than the upgraded M1A2 tanks. However, not everybody thought the US Army needed a new tank at that time, especially the Defense Department, then headed by Secretary of Defense Dick Cheney. This is seen in an extract from a publication by the Historical Office of TACOM titled *Sending the Very Best: An Oral History Interview with Major General Peter M. McVey Program Executive Officer Armored System Modernization*:

> One of the biggest problems we had was with the Defense Department. Defense didn't believe we needed it ... I don't know if he [Cheney] was personally involved or not but he certainly had said: 'We don't need any more tanks.' We took a deep breath and analyzed that and said: 'Well, he didn't say we couldn't improve the ones that we had.' I guess that's how the upgrade program [M1A2 tank] was officially launched.

Rather than building brand new M1A2 tanks from the ground up, all were converted from M1 tanks taken out of storage. As test results from the first 62 units of the M1A2 tank were generally positive, in September 1994, the US Army had another 206 M1 tanks converted into the M1A2 tank by GDLS. By 2001, the US Army had 566 M1A2 tanks in service, all converted M1 tanks. Like the M1A1 AIM and M1A1 SA tank the M1A2 was equipped with both FBCB2 and BFT.

M1A2 Improvements over the M1A1 Tank Series

The M1A2 tank featured a superior level of protection than that found on the M1A1 HA+ tank. It did retain the 120mm main gun fitted to the M1A1 HA+, as well as the earlier tank's secondary armament of three machine guns. However, the effectiveness of the tank's main gun was dramatically improved due to the addition of what is commonly referred to as a 'hunter-killer' sighting arrangement, labeled by the US Army as the Commander's Independent Thermal Viewer (CITV).

The CITV on the M1A2 tank allows the tank commander to search out new targets on his own, while his gunner is engaging a target with his own sights, labeled the Gunner Primary Sight (GPS). It is not to be confused with the global positioning system (which shares the acronym). Upon identifying another more dangerous target with the CITV, the M1A2 tank commander pushes a button that automatically rotates the tank's turret in the direction of the new target so he or the gunner can engage this new more deadly threat. According to test results, it was determined that the addition of the CITV on the M1A2 tank provided for 45 per cent faster target acquisition time than possible with the M1A1 series tank.

On all the various versions of the M1A1 series tanks the tank commander can also take over control of the turret of his tank if he identifies a more hostile threat to his vehicle that the gunner cannot identify, or to assist the gunner in identifying a potential target. He traverses the turret with the Commander's Control Handle Assembly (CCHA) and uses the Commander's GPS Extension to aim the main gun and uses the CCHA to fire the selected weapon system.

Other M1A2 Identifying Features

A second less noticeable change to the external appearance of the M1A2 tank occurred at the tank commander's station on the roof of the vehicle. It was fitted with fewer but larger vision blocks, providing the tank commander with a wider field of view than possible with the smaller vision blocks found on all the earlier version of the Abrams tank series.

On all the Abrams tanks, from M1 to M1A1, not including the M1A1 SA or latest version of the US Marine Corps M1A1 tank, the tank commander's original external cupola design was labeled as the commander's weapon station (CWS). It was in effect an electrically-powered sub-turret that could be traversed 360 degrees to direct the fire of the attached .50 caliber M2HB machine gun. It could also be turned manually if need be.

The tank commander could in theory aim and fire the machine gun mounted on the CWS from within or outside the protection of his tank. The biggest disadvantage with the CWS on the M1 through most versions of the M1A1 tank was the fact that it was not stabilized. This meant that accurately aiming and firing the attached machine gun when traveling over uneven terrain was almost impossible.

The ICWS (Improved Commander's Weapon Station) on the M1A2 tank was different, as it was now fixed in position. The attached .50 caliber M2HB machine gun could be traversed along a 360 degree ring around the ICWS. However, this could only be done manually by the tank commander, which involved projecting his head and upper torso out over the top of his ICWS. Like the original CWS the ICWS is not stabilized and extremely difficult to accurately aim and fire on the move.

Internal Improvements to the M1A2 Tank

A key interior difference between the M1A2 tank and its predecessors was the incorporation of digital technology into the vehicle. The M1A2 tank was 90 percent digital and 10 percent analog, whereas the M1A1 tank was just the opposite, 90 percent analog and 10 percent digital. The advanced digital electronic sensors and systems in the M1A2 tank improved driving, target identification, and the information flow between the computer-driven subsystems and crew.

A description of what the M1A2 tank brought to the US Army is provided by Colonel Chris Cardine (retired), who served as the Abrams series tank program manager from 1994 to 1998:

> The installation of a microprocessor, data bus, monochromatic flat-panel display, and modem in the M1A2 tank opened the floodgates of the information revolution in the army. The tank turned out to be a computer that not only controlled its internal operating functions, but it was also networked. The vehicle could talk digitally to other tanks and computers through digital radios in a local area network. Along with providing protection, mobility, and lethal weapons for the crew, it became in effect their personal digital assistant.

Backdating M1A2 Features to the M1A1

In an early effort at applying some of the advantages offered by the digital infrastructure applied to the M1A2 tank to its M1A1 tank inventory, in 2000, the US Army began fielding an upgraded M1A1 HA tank that was now labeled the M1A1D tank. It was nicknamed the 'Delta' by those who served on them. The letter 'D' stood for digitized. Due to funding constraints, the US Army was not able to afford more than 108 units of the M1A1D tank before the program ended in 2003. In 2004, the US Army pulled all of its M1A1D tanks from service and placed them into storage.

The Evolution of the M1A2 Tank Continues

Even before the last M1A2 tank entered US Army service in 2001 it was decided that its internal electronic hardware and software had to be quickly upgraded to take advantage of new developments in the civilian field, lest it be quickly rendered obsolete.

In September 1994, the US Army awarded a contract to GDLS to begin development of an improved M1A2 tank. This eventually resulted in the creation of the M1A2 SEP tank. The letters 'SEP' stand for system enhancement package. The US

Army manuals for the vehicle list it as Tank, Combat, Full Tracked, 120mm Gun, M1A2 System Enhancement Package SEP, General Abrams.

The first M1A2 SEP tanks entered US Army service in 2001. By December 2004, almost the entire M1A2 tank inventory had been upgraded to the M1A2 SEP tank standard, including approximately 400 M1A1 AIM tanks. The internal electronic upgrades to the vehicle included color display units. It also featured an armor upgrade. Other improvements to the M1A2 SEP included a new more powerful CITV, which allowed the vehicle to accurately engage targets at ranges of over 2 miles.

The first generation M1A2 tank and the improved M1A2 SEP tank were sometimes nicknamed the 'A-deuce' by those who served on them. Some tankers referred to their M1A2 SEP tanks as 'SEPs.' At least one tanker remembers the instructors at the former US Army Armor School at Fort Knox, Kentucky referring to those who were serving on the first unit of M1A2 tanks as 'Jedi Tankers.' This was because they could do everything so much faster and better than those who were still serving on the M1A1 tanks.

Reflecting the advanced capabilities of the entire Abrams tank series compared to US Army tanks that came before, such as the M48 or M60 series, those who served on those older generation tanks are sometimes referred to in good-natured humor as 'Dino Tankers' by Abrams tankers, or the 'old school guys.'

M1A2 SEP Tank Spotting Feature

The one major external difference between the M1A2 and the M1A2 SEP is visible in their respective turret bustles. The M1A2 had a diesel-powered EAPU in its turret bustle. On the M1A2 SEP tank, the diesel-powered EAPU disappeared from the turret basket. In its place appeared a much smaller air conditioning unit labeled the thermal management system (TMS).

The M1A2 SEP was originally intended to be fitted with an under armor gas turbine APU. It was to be installed inside an enclosed compartment on the left side of the engine compartment. However, cost overruns doomed the program and the M1A2 SEP did without any kind of APU.

Improving the M1A2 SEP Tank

The US Army had originally intended that the entire M1A2 SEP fleet would gradually be replaced in service beginning in 2015 by a new family of both manned and unmanned vehicles assigned the name Future Combat System (FCS). Continuing design problems and cost overruns with the FCS pushed the US Army to continue the upgrading of the M1A2 SEP as a fallback position. The FCS program would be abandoned by the US Army in 2009.

In late 2006 the US Army had awarded a contract to GDLS to begin work on an improved M1A2 SEP. That vehicle would be labeled the M1A2 SEP V2. The 'V2' designation stood for Version Two. Upon the introduction of the first M1A2 SEP V2 into US Army service in 2008, the M1A2 SEP was re-designated as the M1A2 SEP V1, to maintain a sense of acquisition sequence.

The M1A2 SEP V2 version of the Abrams tank series is intended to be the cutting edge of US Army ground power until 2024. By 2013, all eleven of the US Army's Armor Brigade Combat Teams (ABCTs) were equipped with the M1A2 SEP V2. One of the US Army National Guard's seven ABCTs is also equipped with the M1A2 SEP V2.

More Upgrades for the M1A2 SEP V2

Unlike the original M1A2 tank model that was 90 percent digital and 10 percent analog, the entire electronic infrastructure of the M1A2 SEP V2 is an open digital architecture designed to accept spin-off technologies without the need for significant re-design. The tank's improved digital displays provide the crew with a better understanding of their vehicle's operational status and their situation on the battlefield.

The US Army's improvement programs for the Abrams series of tank is explained in this quote from an article in the 2014 January–February issue of *Armor* magazine titled *The Armored Brigade Combat Team 2014–2024: Improving Abrams Lethality*, by US Army Major Robert Brown:

> The army's strategy for modernizing the Abrams fleet revolves around incrementally upgrading aspects of the platform through a combination of technological insertion and product improvements based on evolving threats and available technologies ... Recent and continued upgrades to the Abrams MBT will ensure the armored force maintains overmatch and battlefield dominance for the near future ...

The program that Major Brown refers to began in 2012 and is officially referred to as Abrams Engineering Change Proposal, or ECP1. It is intended to redesign and modernize as much as possible of the M1A2 SEP V2 internal features, without degrading any of its current operational performance. The first units of the M1A2 SEP V2 fitted with the ECP1 appeared in 2014.

From a news release issued by Bill Good, PEO Ground Combat Systems, Public Affairs, in October 2012, appears this quote by US Army Lieutenant Colonel William Brennan, product manager for the Abrams, on one aspect of the tank's design that the ECP1 addresses:

> Right now the electrical power is in short supply on the tank. The centerpiece of the ECP1 upgrade will be to restore lost power margin through the integration of a larger generator, improved slip ring, battery management system and a new power generation and distribution system.

Other features of ECP1 include an armor upgrade, an ammunition data link connecting the M1A2 SEP V2 fire control system to the main gun, and a new (un-described) APU for use when the tank's gas turbine engine is off, to power the tank's various auxiliary systems, such as its radio, and electronically powered sighting systems. With ECP1 the current digital single-channel ground and airborne radio system (SINCGARS) is replaced by the joint tactical radio system (JTRS). Unlike the single channel SINCGARS the JTRS is a multi-channel radio that can provide secure voice, video, and data.

Prior to SINCGARS the M1A1 series tanks, except for the M1A1 SA that also had been fitted with SINCGARS, relied on incrementally improved continuously variable FM analog radios, which were very vulnerable to enemy interception or jamming.

Defensive Upgrades with ECP1

Another part of ECP1 for the M1A2 SEP V2 is an updated model of the Counter Remote-Control Improvised Explosive Device (RCIED) Electronic Warfare (CREW) system. CREW is a jamming device that was placed into service by the American military prior to Operation Iraqi Freedom and saw productive use in the subsequent Iraqi Insurgency. It disrupted the radio-frequency (RF) signals sent by Iraqi insurgents over their wireless phones to detonate Remote Control Improvised Explosive Devices (RCIEDs) upon the approach of American soldiers and vehicles. The original version was labeled CREW 1 and the latest version that is fitted to the M1A2 SEP V2, according to the manufacture's website, is designated the CREW 2.

With the successful incorporation of the ECP1 into the M1A2 SEP V2 the US Army is keen to begin the next step, labeled ECP1B. Reflecting the advent of ECP1B, ECP1 has been relabeled ECP1A. One of the key features of the ECP1B when fielded will be the addition of an Active Protection System (APS), which will engage and destroy incoming enemy antitank munitions in mid-air, prior to actually impacting on the tank. Long tested by the American military, up until ECP1B nobody has been willing to embrace its use for fear that when in operation it could pose a danger to nearby friendly personnel.

M1A2 SEP V2 Identifying Features

An external identifying feature seen on many of the M1A2 SEP V2 tanks currently in service, is the extremely large and tall remote-control machine gun mount fitted to the top of the gunner's primary sight (GPS), just in front of the ICWS. It is referred to as the Common Remotely Operated Weapon System (CROWS) and is armed with a .50 Caliber M2HB machine gun. In the near future, the US Army hopes to replace the existing CROWS with a redesigned shorter model that is currently being referred to as the Low Profile CROWS (LP CROWS).

The CROWS is stabilized (unlike the earlier CWS or the later ICWS) and is aimed and fired by the M1A2 SEP V2 tank commander from within the vehicle using a joystick and monitor, being fed video by cameras affixed to the CROWS. These cameras include both a daytime video camera and a thermal imaging camera. To assist the M1A2 SEP V2 tank commander in acquiring the correct range to the target(s) it is also fitted with a laser rangefinder.

CROWS is dismountable and may not always be present on the M1A2 SEP V2. When CROWS is not fitted to the M1A2 SEP V2 the .50 caliber M2HB is refitted to the ICWS. When the CROWS is not apparent on the M1A2 SEP V2 there are three external spotting features to identify it from the earlier and now disappeared M1A2 SEP V1 tank, all of which were converted to the M1A2 SEP V2 standard. These include the following on the rear face of the engine compartment: a Tank-Infantry Phone (TIP) and a Thermal Driver's Rear-View Camera (DRVC) mounted inside the right rear taillight fixture. On the left side of the turret bustle is a small armored box, with a vertical stalk similar in appearance to the crosswind sensor, which is the CREW 2.

Additional External Changes

Other less noticeable changes to the exterior of the M1A2 SEP V2 tank include a second smaller turret basket extension attached to the original turret basket. The second smaller turret basket extension has also been applied to the National Guard's M1A1 SA tanks as well as the US Marine Corps' M1A1 Common tanks.

An important internal addition to the M1A2 SEP V2 tank is located in an enclosed upper portion of the left hull, next to the engine compartment, formerly the location of a fuel tank. It is a set of six heavy duty batteries that now act as the vehicle's APU. This was a feature the M1A2 SEP V1 tank had to do without because of funding constraints. The standard battery arrangement on all the Abrams series tanks, including the M1A2 SEP V2, consists of six heavy-duty batteries located in an enclosed compartment on the right side of the upper hull, next to the engine compartment.

Past the M1A2 SEP V2

The US Army is always looking down the line for its next generation of armored ground combat vehicles. However, for cost reasons the service has realized that a brand new tank is not a realistic option. This is reflected in a US Army News Service article released on 20 October 2014, written by David Vergun, which quotes Brigadier General David Bassett, the program executive officer for ground combat systems. The US Army made the 'difficult decision that it could do more good investing across the entire formation [of ground combat vehicles] rather than a specific vehicle'. Bassett added that the current priority is restoring the armored brigade combat team to its 'relevancy so it can fight with all of the platforms across the entire formation'.

Foreign Users of the Abrams Tank Series

- The Egyptian government asked the American government for permission to license build the M1A1 tank in 1988. Their initial request was for 555 units. Their M1A1 tank fleet has now grown to over 1,000 vehicles. The majority of the components for these tanks came from the United States and were then shipped to Egypt and assembled with some components made in that country.
- Saudi Arabia was given permission by the American government to buy 315 brand new M1A2 tanks from GDLS in 1992, following Operation Desert Storm, at the same time the US Army was taking them into service. These were delivered between 1993 and 1996.
- Not wanting their M1A2 tanks to become obsolete, the Saudi government contracted with GDLS in 2006 to rebuild and upgrade their M1A2 tank fleet to the US Army M1A2 SEP standard, with some modifications to suit the needs of the Royal Saudi Army. These upgraded Saudi M1A2 tanks are referred to by GDLS as the M1A2S.
- In 2006, the Saudi government also received permission to buy fifty-eight M1A1 tanks from the US Army storage stockpile. These M1A1 tanks were put through the AIM program and then upgraded by GDLS to the M1A2S tank standard.
- The government of Kuwait was given permission by the American government to buy 218 brand-new M1A2 tanks, which were delivered between 1994 and 1997. There has been no effort on the part of the Kuwait government to have those tanks rebuilt or upgraded.
- In 2005, the Australian government ordered fifty-nine M1A1 SA tanks for its army, as a replacement for their existing fleet of German-designed and built Leopard I tanks. The M1A1 SA tanks for the Australian Army came from the US Army M1A1 storage stockpile, and were put through the AIM program before being delivered in 2006.
- To bolster the strength of the American-trained and supplied Iraqi Army, the US government provided 140 M1A1 SA tanks in 2008, under a foreign military sales agreement. The first of these tanks, all having been put through the AIM program, were delivered in 2010, the last in 2011. It is unknown what level of armor protection the tanks had.
- With the disintegration of the Iraqi Army, under attack from the Islamic State of Iraq (ISIL) in June 2014, a number of the American-supplied M1A1 SA tanks were destroyed and some captured intact. The ability of the ISIL fighters to successfully operate the M1A1 SA tanks they captured is doubtful. In an attempt to re-build the Iraqi Army the second time the American government sought to provide what was left of it with a new fleet of M1A1 SA tanks in late 2014.

(*Above*) A key identifying feature on the M1A2 tank to distinguish it from the M1A1 tank, as they both were fitted with the same 120mm main gun, is the Commander's Independent Thermal Viewer (CITV), the upper portion of which projects out of the turret roof, just in front of the loader's hatch. The CITV can be seen in this picture of an M1A2 tank. (*GDLS*)

(*Opposite above*) The CITV was originally envisioned by the US Army for fitting to the M1A1 tank, but cost overruns pushed that back until the M1A2 tank could be fielded. The CITV seen here on this M1A2 tank at Fort Knox, Kentucky, provides the tank commander with a backup sight having the same capabilities to that of the Gunner's Primary Sight (GPS). (*GDLS*)

(*Opposite below*) The various external and internal improvements made to the M1A2 tank, which greatly increased its combat capabilities are listed in the line drawing. One of the key internal changes to the M1A2 tank was the incorporation of an electronic device known as the Inter-Vehicular Information System (IVIS). Connected to the IVIS was a position/navigation (POS/NAV) system. (*DOD*)

"The First Information Age System"
M1A2

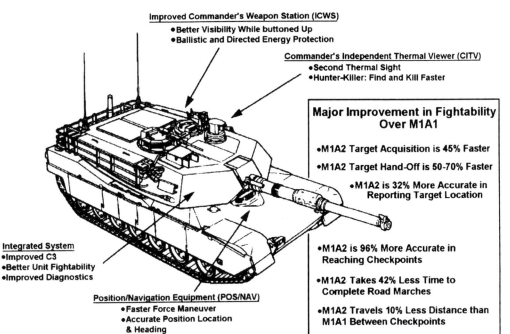

Improved Commander's Weapon Station (ICWS)
- Better Visibility While buttoned Up
- Ballistic and Directed Energy Protection

Commander's Independent Thermal Viewer (CITV)
- Second Thermal Sight
- Hunter-Killer: Find and Kill Faster

Integrated System
- Improved C3
- Better Unit Fightability
- Improved Diagnostics

Position/Navigation Equipment (POS/NAV)
- Faster Force Maneuver
- Accurate Position Location & Heading

Major Improvement in Fightability Over M1A1

- M1A2 Target Acquisition is **45% Faster**
- M1A2 Target Hand-Off is **50-70% Faster**
- M1A2 is **32% More Accurate** in Reporting Target Location
- M1A2 is **96% More Accurate** in Reaching Checkpoints
- M1A2 Takes **42% Less Time** to Complete Road Marches
- M1A2 Travels **10% Less Distance** than M1A1 Between Checkpoints

M1A2 ABRAMS TANK SYSTEM
TURRET INTERIOR

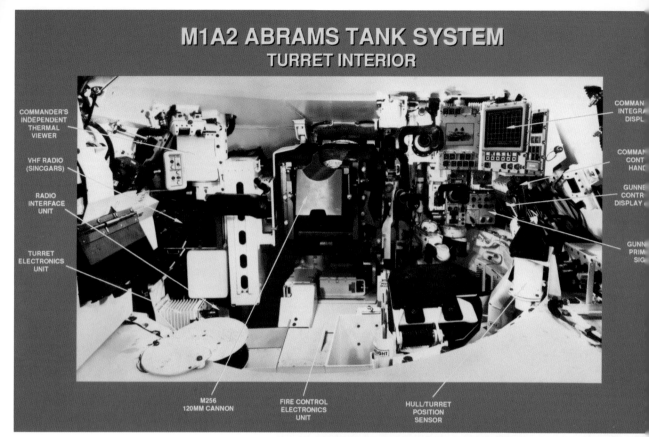

- COMMANDER'S INDEPENDENT THERMAL VIEWER
- VHF RADIO (SINCGARS)
- RADIO INTERFACE UNIT
- TURRET ELECTRONICS UNIT
- M256 120MM CANNON
- FIRE CONTROL ELECTRONICS UNIT
- HULL/TURRET POSITION SENSOR
- COMMAN[DER'S] INTEGRA[TED] DISPL[AY]
- COMMAN[DER'S] CONT[ROL] HAND[LE]
- GUNNE[R'S] CONTR[OL] DISPLAY
- GUNN[ER'S] PRIM[ARY] SIG[HT]

Visible in this picture are various features found inside the M1A2 tank turret. The Commander's Integrated Display (CID) screen, part of the Inter-Vehicular Information System (IVIS), is seen on the right of the photograph and is labeled. To the left of the CID is the video feed from the Commander's Independent Thermal Viewer (CITV). (*GDLS*)

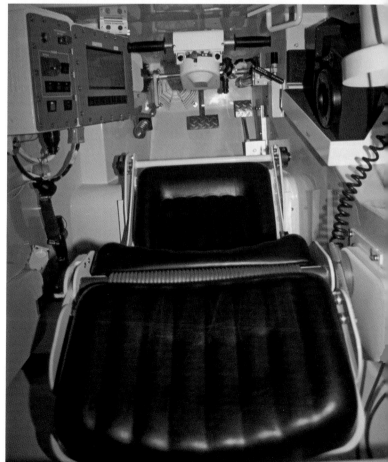

The driver's station on an M1A2 tank training simulator is shown. Rather than depend solely on voice commands to direct his driver around the battlefield the M1A2 tank commander can direct him via a driver's instrument display (DID) panel. The DID is located to the left of the driver's seat and seen on this driving simulator. (*Michael Green*)

Another external feature of the M1A2 tank that differs from the M1A1 tank was a newly-designed tank commander cupola seen here. On the M1A2 the tank commander's station is a non-rotating fixture with a manually-operated .50 caliber M2HB machine gun set on a pedestal that rotates around it. It goes by the name Improved Commander Weapon Station (ICWS). On the M1 through M1A1 tank series the entire cupola, designated the Commander Weapon Station (CWS), turned with its attached machine gun. *(Michael Green)*

This close-up picture shows the bottom portion of the pedestal that supports the .50 caliber M2HB machine gun that in turn is attached to the Improved Commander Weapon Station (ICWS) on the M1A2 SEP tank. Despite it being in service since before the Second World War the M2HB machine gun remains a fearsome weapon for all those who go up against it in battle. *(Brent Sauer)*

Pictured here is the Army's Embedded Global Positioning System Receiver (AEGR) antenna mounted on the roof of the M1A2 tank. It is located behind the Commander's Independent Thermal Viewer (CITV) for protection. Along with the Position/Navigation Unit (POS/NAV) the AEGR provides the M1A2 tank enhanced navigational capabilities. *(Michael Green)*

This line drawing shows the many improvements made to the M1A2 SEP tank. As labeled in the drawing, the M1A2 SEP had a much more powerful thermal sight installed within the Commander's Independent Thermal View (CITV). The printed manuals familiar to generations of US Army tankers were replaced by digital manuals embedded in the tank's computer, with the introduction of the M1A2 tank. *(DOD)*

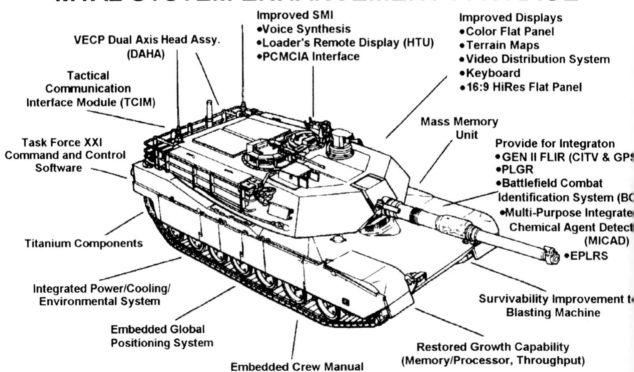

M1A2 SYSTEM ENHANCEMENT PACKAGE

- VECP Dual Axis Head Assy. (DAHA)
- Tactical Communication Interface Module (TCIM)
- Task Force XXI Command and Control Software
- Titanium Components
- Integrated Power/Cooling/Environmental System
- Embedded Global Positioning System
- Embedded Crew Manual
- Improved SMI
  - Voice Synthesis
  - Loader's Remote Display (HTU)
  - PCMCIA Interface
- Improved Displays
  - Color Flat Panel
  - Terrain Maps
  - Video Distribution System
  - Keyboard
  - 16:9 HiRes Flat Panel
- Mass Memory Unit
- Provide for Integraton
  - GEN II FLIR (CITV & GPS)
  - PLGR
  - Battlefield Combat Identification System (BC
  - Multi-Purpose Integrate Chemical Agent Detect (MICAD)
  - EPLRS
- Survivability Improvement t Blasting Machine
- Restored Growth Capability (Memory/Processor, Throughput)

Seen here is a close-up view of the new rear engine compartment grill arrangement of an M1A2 SEP tank. The center section grill is for the engine exhaust. The black paint is a special formula intended not to peel off under the intense engine exhaust heat. Its lacks the multi-hole bracket surrounding the center section engine exhaust grill, which was the attachment point for the lower portion of the vertical fording tube as used on USMC M1A1 Common tanks. *(Michael Green)*

In front of the tank commander of this M1A2 SEP tank is the new two-screen Commander's Display Unit (CDU). It was the replacement for the Commander's Integrated Display (CID) screen, part of the Inter-Vehicular Information System (IVIS), on the M1A2 tank. The upper screen is the video feed from the Commander's Independent Thermal View (CITV). *(DOD)*

An external change to the M1A2 SEP tank was the removal of the large diesel-powered Auxiliary Power Unit (APU) in the turret bustle. It was replaced on an M1A2 SEP tank by a smaller vapor-compression system unit (VCSU), as seen here, located on the left side of the turret bustle. The VCSU forms the upper portion of the tank's thermal management system (TMS). In civilian terms the TMS is an air-conditioning system. *(Michael Green)*

The vapor-compression system unit (VCSU) of an M1A2 SEP tank in Iraq can be seen in its turret basket. Notice the heat shield affixed to the tank's rear engine exhaust grill. The vehicle's loader is cautiously peering out over the lip of his overhead hatch, which indicates that they have recently taken enemy fire or anticipate being fired upon. *(DOD)*

Two US Army M1A2 SEP tanks are shown on patrol in a war-ravaged Iraqi town. The M1 up through most of the M1A1 series tanks were equipped with FM analog voice radios. The M1A2 series tanks are equipped with a digital radio referred to as the Single-Channel Ground and Airborne Radio System (SINCGARS). The new digital radio first entered US Army service in 1990. *(DOD)*

(*Above*) Notice the tow bar attached to the front hull of the M1A2 SEP tank in Iraq. This feature was often seen on Abrams tanks in Iraq and reflected the desire by the crews of the vehicles to be able to tow each other out of harm's way without waiting for dedicated armored recovery vehicles (ARV), such as the M88A2, which were often in short supply. (*DOD*)

(*Opposite above*) If one wonders just how much gear that a crew of an M1A2 SEP tank can jam into the two turret bustles of their tank, this picture taken in Iraq in 2004 will answer that question. Unfortunately, when in combat, it's the personal gear of the crew that is often riddled by bullets and assorted other battlefield fragments, making life in the field just a bit more miserable. (*Brent Sauer*)

(*Opposite below*) In this picture we see the two blow-off panels located on the roof of the rear turret bustle of an M1A2 SEP tank. Notice that the small adjusting wheels seen on US Army and US Marine Corps M1A1 series tanks built after November 1990 are no longer present. However, the cut-out locations for the fitting of the adjusting wheels remain on the blow-off panels. (*Michael Green*)

(*Above*) With the advent of an upgraded model of the M1A2 SEP tank, referred to as the M1A2 SEP V2 as seen here the M1A2 SEP tank was relabeled as the M1A2 SEP V1. Both versions would see combat in Iraq. Eventually all the M1A2 SEP V1 tanks in the US Army's inventory were upgraded to the M1A2 SEP V2 standard. (*USABOT*)

(*Opposite above*) This M1A2 SEP must have just rolled off the factory floor as it has not yet been fitted with the standard two radio antennas. The flat panels on either side of the front of the vehicle's turret, and the ribbed devices on the side of the turret, are Identification Friend or Foe (IFF) panels. The American military official name for them is Combat Identification Panels (CIP). Projecting out from the right hand side of the turret is a Blue Force Tracking (BFT) unit with a white plastic cover. (*DOD*)

(*Opposite below*) Pictured during a training exercise is a US Army M1A2 SEP V2 tank. Notice the two radio antennas, one on either side of the vehicle's rear turret bustle. One is the transmitter antenna and the other is the receiver antenna. The Single-Channel Ground and Airborne Radio System (SINCGARS) fitted to the M1A2 series of tank can accommodate voice, analog and digital data transmissions. (*DOD*)

(Above) From combat experience gained during the fierce and often close-in fighting of the Iraqi Insurgency, the US Army saw a need to field a weapon system that would allow the tank commander of the M1A2 SEP V2 to accurately engage enemy dismounts with his .50 caliber M2HB machine gun from within his tank. The answer was the Common Remotely Operated Weapon Station (CROWS) seen here on the turret roof of the vehicle pictured. *(DOD)*

(Opposite above) Taking part in a training exercise at the National Training Center (NTC), located at Fort Irwin, California, is this M1A2 SEP V2 tank. One can make out the Common Remotely Operated Weapon Station (CROWS) on the cluttered turret roof of the vehicle. The black multi-slotted device affixed to the front of the tank's turret is Main Gun Signature Simulator (MGSS), which forms part of the Multiple Laser Engagement System (MILES). *(DOD)*

(Opposite below) Taken during a pause in a training exercise is an M1A2 SEP V2 as is indicated by the Common Remotely Operated Weapon Station (CROWS) on the turret roof. The CROWS is mounted on a removable metal bracket seen in this picture that covers the armored housing over that portion of the Gunner Primary Sight (GPS) that projects out over the top of the turret roof. *(DOD)*

(*Above*) The Common Remotely Operated Weapon Station (CROWS) seen here on this M1A2 SEP V2 is missing its .50 caliber M2HB machine gun. With the machine gun fitted the CROWS weighs 430lbs. It can only be installed or removed with an overhead crane. The ammunition container of the CROWS holds 400 rounds of .50 caliber ammunition. (*James Warford*)

(*Opposite above*) A view of the bracket that supports the Common Remotely Operated Weapon Station (CROWS) is seen on an M1A2 SEP V2 taking part in a training exercise. On the rear of the turret can be seen the projecting platform for the Blue Force Tracking (BFT) unit with a black plastic cover. This location would better protect it from the muzzle blast of the CROWS .50 caliber M2HB machine gun. (*DOD*)

(*Opposite below*) From a different angle is this picture of an M1A2 SEP V2 during a training exercise. No doubt extremely useful in urban combat the Common Remotely Operated Weapon Station (CROWS) seen on this vehicle does block the forward vision of the tank commander when he raises his head out over the top of his Improved Commander's Weapon Station (ICWS). (*DOD*)

Being guided onto a railcar is an M1A2 SEP V2 tank. Notice that the new second smaller turret basket extension, which now appears on all current models of the Abrams series tanks in service, can be collapsed. The vertical stalk, with the flat top, seen on the right side of the main turret bustle is a new electronic countermeasure (jamming) device, referred to by the builder as the CREW Duke System. *(DOD)*

From the loader's position on an M1A2 SEP V2 tank can be seen the right side of the turret crewed by the tank commander. Visible in front of the tank commander is the two-screen Commander's Display Unit (CDU). Just below the CDU is the Gunner's Control and Display Panel (GCDP). *(DOD)*

Caught at the moment of firing is an M1A2 SEP V2 tank. Visible on the rear of the vehicle's engine compartment are its two tail light assemblies. The one on the right hand side contains a camera for the driver and is labeled the Thermal Driver's Rear-view Camera (DRVC). To the left of that taillight are two armored boxes, the larger of which contains the Tank Infantry Phone (TIP). *(DOD)*

Moving to a training exercise area in Germany is an M1A2 SEP V2 as indicated by the Common Remotely Operated Weapon Station (CROWS) on the turret roof of the vehicle pictured. The CROWS can be traversed 360 degrees, has an elevation of 60 degrees and can be depressed 20 degrees. It is equipped with both a daylight camera as well as a thermal camera, both having an auto focus feature. *(DOD)*

Seen on a firing range are two M1A2 SEP V2 fitted with the Common Remotely Operated Weapon Station (CROWS). American Abrams tankers are taught in training to constantly scan their assigned areas of observation at all times to detect targets or target signatures, such as black plumes of exhaust smoke from a diesel tank engine. This is to be done with the vehicle's optics, binoculars, and the unaided eye. *(DOD)*

Chapter Four

Marine Corps Abrams Tanks

The US Marine Corps (USMC) began taking a serious look at the ever evolving threat posed by new generations of Soviet Army MBTs in the late 1970s. At that point, it was clear to many that their inventory of M60A1 tanks was obsolete. Some thought was given to a new and improved version of the tank. However, as Marine tankers depended on logistics support for their vehicles from the US Army, and that service was switching from the M60 series of tanks to the Abrams tank series, it was clear that it was not a workable solution.

Hobbled by budget shortfalls, the idea of acquiring the M1A1 tank appeared and disappeared from the USMC budget beginning in 1981. Finally, in February 1985, the Commandant of the Corps set an objective of acquiring 490 M1A1 tanks to replace the 760 M60A1 tanks then in the inventory. Plans called for the first M1A1 HA tanks to be delivered by November 1990, with all remaining vehicles being delivered by the end of 1992. Due to some senior level miscalculations, the USMC would only receive 269 units of the M1A1 HA tank, out of the 490 originally approved by Congress.

Design Changes
The 269 brand new M1A1 HA tanks that came off the GDLS production lines, beginning in November 1990 for the USMC, as well as all brand new US Army M1A1 HA tanks produced after that date differed from the M1A1 HA tanks built for the US Army since May 1988. These differences consisted of eighty engineering changes. Three of these changes were intended for the M1A1 HA tanks being built for the USMC. They included a deep water fording kit, stronger chain tie-downs, and a position location reference system, none of which the US Army adopted.

Reflecting all the design features now shared by the two services' M1A1 HA tanks built after November 1990, the unofficial term 'common tank change' came into use. US Army and USMC tankers shorten that to just the word 'common.' Hence, you have some US Army tankers in the early 1990s referring to their post-November 1990 built M1A1 HA tanks as the M1A1 Heavy Common (HC) tank. This was due to its DU armor package and the fact that it was almost identical to the M1A1 HA tanks then being supplied to the USMC. Marine tankers labeled their Abrams tanks as the 'M1A1 Common tank.'

Congress Comes To the Rescue

To help the often cash-strapped USMC, Congress provided funding in 1994 and 1995 to arrange for the transfer of 134 M1A1 tanks from the US Army inventory into theirs. These vehicles were transferred from a US Army tank division stationed at Fort Hood, Texas that was transitioning into the M1A2 tank.

The M1A1 tanks the US Army was turning over to the USMC in the mid-1990s were all built prior to November 1990 and therefore lacked the eighty engineering changes that came with the common tank change. An unofficial USMC label for the pre-November 1990 built M1A1 HA tanks was 'Plain Janes.'

To bring these Plain Jane M1A1 HA tanks up to the same standard as the post-November 1990 M1A1 Common tanks already in the USMC inventory, Lieutenant Colonel Dennis Beal, the USMC Tank Program Manager, developed an upgrade program. This upgrade program was applied at Anniston Army depot between 1995 and 1997 at a cost of 41 million dollars.

Between 2002 and 2005, the US Army transferred an additional 181 M1A1 HA tanks to the USMC. These needed minor applications and modifications to bring them to the current M1A1 Common configuration. Most of these vehicles came from the US Army National Guard and were modified before delivery.

US Marine Corps M1A1 Tanks in Combat

To assist the USMC during Operation Desert Storm in 1991, since it had not completely switched over from the M60A1 tank to the M1A1 Common tank, the US Army loaned it M1A1 HA tanks from its own inventory. None of the loaned Abrams tanks employed by the USMC during the brief ground campaign phase of Operation Desert Storm where struck by enemy fire.

During a night-time combat engagement during the ground warfare phase of Operation Desert Storm, an elite Iraqi Republican Guard T-72 equipped tank battalion blundered into a parked US Marine Corps reserve tank company. By taking advantage of their M1A1 HA tank's thermal imaging gun sights, which the enemy vehicles lacked, they opened fire first and destroyed thirty-four of the thirty-five Iraqi tanks in approximately ninety seconds.

A feature seen mounted on the front turret roof of some of the US Army supplied M1A1 HA tanks during Operation Desert Storm was a small device in front of the loader's overhead hatch that looked like a television set and was designated the Missile Countermeasure Device (MCD) AN/VLQ-6. The MCD was a multi-threat jammer intended to protect the vehicle it was mounted on against a wide range of ground and air-launched Anti-Tank Guided Missile (ATGM) threats. It was not mounted on US Army M1A1 HA tanks during Operation Desert Storm, but did appear on the US Army's fleet of M2A2 Bradley Infantry Fighting Vehicles (IFVs) during the conflict. The mounting rails for the device still appear on some USMC M1A1 Common tanks.

Iraq and Afghanistan

As in Operation Desert Storm, the USMC Abrams tanks also performed exceptionally well in combat during Operation Iraqi Freedom and in the fighting that following. Some service observations on the performance of the tank can be found online in a PowerPoint presentation by Lieutenant Colonel Skip Gaskill, the Marine Program Manager, Tank Systems. In that presentation Gaskill mentions the following under a heading titled 'Survivability': 'Frontal armor and [armored side] skirts as advertised – superior! Individual vehicles took dozens of RPG hits. [Engine] Grill doors as expected, susceptible to RPGs and medium caliber weapon. Mobility kills.'

Unlike the US Army, that never deployed any of its Abrams tanks to Afghanistan, the US Marine Corps decided in the fall of 2010 that tanks could serve a useful role in aiding Marine infantrymen in taking on the Taliban. A company of fourteen M1A1 Common tanks was eventually deployed to the country in 2011. They were withdrawn in late 2013.

An external identifying feature on the USMC Abrams tanks that served in Iraq from 2003 on, and in Afghanistan, were two large antennas at the rear of their turrets that marked them as being fitted with the CREWS 1 electronic receiver/jamming unit. From the manufacture's website is this extract describing the device: 'The IED-jammer has a wide receive and transmit frequency coverage and handles multiple threats simultaneously.' The designer and builders of the device state that it can also easily be reprogramed to meet new threats encountered on the battlefield.

Marine Corps Tank Upgrades

The USMC had continuously added various upgrades over the years to their M1A1 Common tanks to keep them viable on the battlefields of the future. The current USMC plans call for keeping them in service until 2050. However, lacking the funding to keep up with the US Army's adoption of the M1A2 series of Abrams tanks has left the USMC inventory of M1A1 Common tanks behind in capabilities.

Current USMC M1A1 Common tanks have some internal features found within the US Army's M1A2 tank fleet. But the most important feature of the M1A2, the CITV, has not been fitted to the Marine M1A1 Common tanks and there are no plans to do so. Instead, the USMC has been making some incremental improvements to their M1A1 Common tank inventory under a program referred to as the Main Battle Tank Fleet [MBT] Modification Kit.

Some of the MBT Modifications Kit Program upgrades such as the SCWS and TIP are the same as those fitted to the US Army National Guard's M1A1 SA tank fleet. As with the US Army's M1A2 tanks and the US National Guard M1A1 SA all the USMC

M1A1 Common tanks have been fitted with a second smaller turret basket extension, attached to the rear of the original tank's turret basket.

Under the MBT Fleet Modification Kit Program the Marines have also made some specific upgrades to their M1A1 Common tanks, which the US Army has not. One of these is referred to as Abrams Suspension Upgrade (ASU) intended to allow the tank's suspension system to support as much as 77 tons of weight.

Another Marine-specific change to their M1A1 Common tanks under the MBT Fleet Modification Kit Program was the revamping of the tank's rear turret bustle main gun ammunition storage compartment. This upgrade was begun in 2014. At the same time, the main gun ammunition storage compartment was being redone. The two existing steel armor blow-off panels, with the adjuster wheels, were replaced by lighter titanium blow-off panels that do not have adjuster wheels fitted.

Despite lacking the funding to upgrade their M1A1 Common tanks with the CITV as found on the US Army M1A2 tanks, the USMC made some upgrades to their vehicle's fire-control system under the MBT Fleet Modification Kit Program. They fell under the heading Firepower Enhancement Program (FEP). From a USMC website appears this passage explaining what the FEP brought to the M1A1 Common tank.

> The Abrams Integrated Display and Targeting System includes a high-resolution thermal camera and a fixed high resolution color display designed to streamline information flow, and serve as the tank's Blue Force [friendly forces] Tracking system display and interface. The Abrams Integrated Display and Targeting System allows the tank commander to accurately detect, recognize, identify, and engage targets under limited visibility conditions and at ranges approaching the full capability of the tank.

Reflecting the FEP upgrade to their M1A1 Common tanks, some Marine tankers have taken to referring to the tank as the M1A1 FEP. Although most often they just refer to it as an Abrams tank.

(*Opposite above*) A USMC M1A1 Common tank is being directed onto a motorized pontoon bridge. Due to its traditional doctrine of lightness, so as not to replicate the US Army, the leadership of the Corps has never fully embraced the need for having large and heavy tanks. But outside events have always pushed them into acquiring them. (*DOD*)

(*Opposite below*) Driving off the ramp of a Joint High Speed Vessel (JHSV) during a test is a USMC M1A1 Common tank. It was Lieutenant Colonel Martin R. Steele who finally convinced the Corps' senior leadership in 1985 of the need to replace the aging inventory of M60A1 tanks, first adopted in 1974, with the new M1A1 Abrams tank being built for the US Army. (*DOD*)

Here we see two USMC M1A1 Common tanks during a training exercise. Visible on either side of the tank's turret in the foreground are six-barrel M250 smoke grenade dischargers. The box in front of the M250 on the right hand side of the picture is for spare smoke grenades. There is a matching box with spare smoke grenades on the opposite side of the turret. (DOD)

Pictured is a USMC M1A1 Common tank traveling at high speed in Iraq. Notice the omnipresent cooler on the rear roof of the vehicle. The effective use of the M250 smoke grenade launcher system in combat depends on the crew's awareness of the wind speed and direction. If the wind is too strong, or blowing in the wrong direction, it can prove difficult to establish a useful smoke screen. (DOD)

The main role for all the tanks that have been part of the USMC inventory is the support of Marine infantrymen, as is seen here in Iraq. All Abrams series tanks, be they USMC or US Army, have been fitted with an automatic fire-suppression system that uses a system of sensors arrayed around the interior of the tank's turret and hull to detect and extinguish fires before they become a threat to the crew. (DOD)

A Marine tanker is seen fueling his M1A1 Common tank. From the M1 through all the various models of the M1A1 tank series the vehicles have had four under-armor fuel tanks that hold a total of 505 gallons of fuel. There are two in the front hull, one on either side of the driver's position; two more are located in the rear of the tank's hull, with one on either side of the engine compartment. (DOD)

(*Above*) With the help of a ground guide ('there should always be two' according to the manuals) the crew of this USMC M1A1 Common tank, fitted with its fording gear, is going to be directed onto the US Navy tank landing craft in the background. Due to the weight of the tank they can only be taken onboard and disembarked in fair weather conditions, in mild surf. (*DOD*)

(*Opposite above*) The two tall vertical stacks of the fording kit on the right side of this USMC M1A1 Common tank cover the air intakes for the vehicle's engine, while the tall single stack at the rear of the engine compartment vents out the engine exhaust gases. These gases are so hot that tank crews are warned about the potential for starting fires on dry grass or shrubbery. (*DOD*)

(*Opposite below*) US Marine Corps M1A1 Common tanks are seen on a stateside firing range. On the rear of the turret bustle is the largest of the three fording tubes, the other two smaller fording tubes are stored within the largest of the three fording tubes. As can be deduced, the reloading of the .50 caliber M2HB machine gun fitted to the Commander Weapon Station (CWS) can be a dicey affair when under enemy fire. (*DOD*)

Shown at the moment of firing on a training range is a USMC M1A1 Common tank. Visible at the rear of the engine compartment is the bottom portion of the engine exhaust fording kit, minus the fording stack. With the External Auxiliary Power Unit (EAPU) taking up much of the space in the tank's original turret bustle the need for an extension shown here to store the crew's personal gear, such as sleeping bags, is clear. *(DOD)*

Visible on this USMC M1A1 Common tank, taking part in an urban training exercise, are the two blow-off panels mounted on the roof of the rear turret bustle. On the top of the blow-off panels are the six small adjusting wheels. Combat has shown that the blow-off panels function as designed and have saved the lives of many an Abrams tank crewman. *(DOD)*

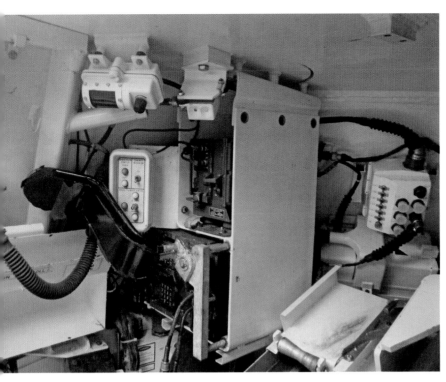

This picture taken from the tank commander's side of a USMC M1A1 Common tank shows the arrangement of the shelving in front of the loader's position. Mounted on the shelving is the tank's radio. On the Abrams series of tanks the coaxial machine gun is located on the gunner's side of the turret. (*Don Moriarty*)

In this picture of a USMC M1A1 Common tank we can see the covered Missile Countermeasure Device (MCD) AN/VLQ-6 mounted on the protrusion seen in front of the loader's hatch, referred to by some as the 'manhole cover.' It was applied to late-production M1A1 tanks for the possible fitting of a second optic sight that never occurred. (*DOD*)

Visible in front of the loader's position on this USMC M1A1 Common tank are the two mounting rails for the fitting of a Missile Countermeasure Device (MCD) AN/VLQ-6. These mounted rails are still seen on some Marine tanks. The original CWS in this picture was at some point updated with a thermal camera fitted to the right of the .50 caliber machine gun, just in front of the ammunition box support platform. *(DOD)*

Visible in this picture of a USMC M1A1 Common tank in Iraq are the two large antennas that are the external spotting features of the Remote-Control Improvised Explosive Device Electronic Warfare (CREW 1) system. On this particular vehicle the CREW 1 dual antennas are mounted in the rear turret bustle extension. *(DOD)*

On this USMC M1A1 Common tank we see the two large external antennas, which form part of the Counter Radio-Control Improvised Explosive Device (RCIED) Electronic Warfare (CREW 1) system. Pictorial evidence seems to indicate that they are typically located together in the rear turret bustle extension of Abrams tanks. On this vehicle, however, one of the two antennas is located on the right front of the vehicle's turret. *(DOD)*

In this picture we see a USMC M1A1 Common tank on patrol in Afghanistan. The USMC had seventeen M1A1 Common tanks flown to that country on US Air Force C-17 Globemaster transport planes. Notice the vehicle's standard two radio antennas, as well as the two large Counter Radio-Control Improvised Explosive Device (RCIED) Electronic Warfare (CREW 1) system antennas. *(DOD)*

(Opposite above) On patrol in Afghanistan are two USMC M1A1 Common tanks. The most effective use of the tanks in that country was employing their long-range observation capability in identifying enemy personnel setting up Improvised Explosive Devices (IEDs) both during daylight hours, but especially at night. The tank's main guns also proved extremely effective in blowing holes in thick compound walls that the enemy employed as cover. *(DOD)*

(Opposite below) This USMC M1A1 Common tank features the Stabilized Commander's Weapon Station (SCWS), also seen on US Army National Guard M1A1 SA tanks. The tan-painted SCWS stands out against the NATO three-color camouflage paint scheme of green, brown, and black, adopted by the American military in 1987, and applied at the factory. *(DOD)*

(Above) An overhead picture of the Stabilized Commander's Weapon Station (SCWS), minus its .50 caliber M2HB machine gun, fitted on a USMC M1A1 Common tank. The stabilization system, which is the heart of the SCWS, is electrically powered. The large rear frame of the device is intended as a counterweight to help balance the stabilization system when in operation. *(Don Moriarty)*

On the move at a stateside training base is a USMC M1A1 Common tank. It is fitted with the Stabilized Commander's Weapon Station (SCWS). The USMC's inventory of Abrams tanks lack the US Army Force XXI Battle Command and Brigade and Below (FBCB2), or Blue Force Tracking (BFT) systems, as do the US Army M1A2 series tanks, or the US National Guard M1A1 SA tanks. (DOD)

Chapter Five

Firepower Close-Up

The 120mm main gun on the M1A1 through M1A2 tanks has an approximate length of 18 feet 4 inches. The gun tube weighs 2,590lbs, the breech mechanism 1,506lbs and the breechblock 100lbs. The gun tube is supported by a gun mount that weighs 7,307lbs, which supports the gun on trunnion bearings installed on pins set in the front of the turret. A hydraulic elevating mechanism allows for the elevation and depression of the main gun. The main gun can be elevated between +20 and −10 degrees from zero (horizontal).

Smoothbore tank guns have some advantages over rifled tank guns. They can withstand higher pressure, are cheaper to build, and do not wear out as quickly. Given the same amount of propellant in a main gun cartridge case, the muzzle velocity of a smoothbore tank gun is higher. The 105mm main guns on the M1 and IPM1 were rifled.

Main Gun Ammunition

American tank main gun ammunition has generally been fixed (one-piece), meaning the cartridge case is crimped to the projectile. The major components of an Abrams series tank's 120mm main gun round are the primer, the propelling charge, and the projectile. The primer receives the electric firing impulse and ignites the propellant contained within the cartridge case. The propellant does not explode, but burns extremely quickly, and pushes the projectile down and out of the barrel.

To aid in the disposal of the large cartridge cases of the 120mm rounds fired by the M1A1 through M1A2 tanks, the cartridge case is semi-combustible, with the only portion of the cartridge case expelled from the gun's breech upon recoil being a small, approximately 10lb steel rubber-tipped stub, officially referred to as the 'case base.' Its popular nickname is 'aft cap.' It is the most common souvenir retained by Abrams tankers upon leaving military service.

Early Objections

One of the original objections to the mounting of the German-designed 120mm main gun in the M1A1 Abrams by many in the US Army armor community was its use of a semi-combustible cartridge case. Prior US Army experience with fully-combustible cartridges cases employed for the 152mm gun/missile launcher mounted in the M60A2 main battle tank and M551 Sheridan light tank had not been positive. Early problems with the German-designed 120mm semi-combustible cartridge case had also worried many even though they were quickly resolved.

A disadvantage with the semi-combustible cartridge case used on the Abrams tank is the propellant contained within, which can be accidentally ignited during loading, with fatal results for the crew. This can happen two ways. First, burning or smoldering residue from the previous round fired can remain in the chamber. When the next round is loaded, the burning residue can ignite the propellant when the breech is open and when the loader is pushing the new round into the chamber. The chemistry of the 120mm semi-combustible cartridge case employed on the M1A1 and M1A2 series tanks is carefully controlled to reduce any chances of burning residue being left behind in the chamber.

The other way the propellant in the current semi-combustible cartridge cases can be accidentally ignited is during a 'flareback.' Propellants do not contain enough oxygen to burn completely and they stop burning when all the oxygen is gone. The remaining high-temperature propellant gas resumes burning when it contacts the air outside the muzzle end of the tank's main gun. If the bore evacuator, a British invention, used to clear fumes and smoldering residue from a gun barrel is not functioning properly, the propellant gas can flare back into the turret when the breech is opened, possibly igniting the next cartridge case. Since the flareback lasts only a fraction of a second, Abrams loaders are taught not to remove the next round or even open the blast-proof doors in the rear turret bustle until the main gun has fired, recoiled, and ejected the case base.

Tank Killing Rounds

The primary tank killing round for the 120mm main guns on the M1A1 and M1A2 tanks has always been a family of Kinetic-Energy (KE) rounds, beginning with the M829 Armor-Piercing, Fin-Stabilized, Discarding Sabot, Tracer (APFSDS-T). The most modern version of this round is designated the M829A4 and entered service in 2014. The letter 'T' in the ammunition designation refers to tracer. In tank fire commands all the various versions of APFSD-T are referred to as 'SABOT' (pronounced SAY-BO).

Like all KE rounds, the M829 series of KE rounds depend on their velocity, mass, hardness, and the length of their penetrator to punch through a tank's armor. All the M829 series penetrators are made of DU which, upon striking an enemy tank, heats to several thousand degrees as it punches its way through. If the DU penetrator successfully passes through a tank's armor, fragments of its own armor and the pyrophoric fragments from what is left of the DU penetrator spew into the targeted vehicle, with disastrous effect on the crew and interior components of the victim tank.

The DU penetrator of the M829 series of KE rounds is centered in a lightweight three-piece, metal alloy body, labeled the 'sabot,' which is crimped to the cartridge case. The three-piece sabot falls away as the projectile exits a tank's barrel. These sabot petals are extremely dangerous once they leave the muzzle of a tank's main gun and tank crews are cautioned not to fire over the heads of friendly troops.

HEAT Rounds

The original chemical energy (CE) main gun round employed by the M1A1 and M1A2 tanks was the M830 HEAT-MP-T. The letter designation code 'HEAT' stands for high-explosive antitank, with the letters 'MP' standing for multipurpose. In tank fire commands, the round is referred to as 'HEAT.' The intended targets for HEAT include lightly armored or non-armored targets, field fortifications, and personnel.

In a book titled *War Stories of the Tankers*, published in 2008, and edited by your author, comes this passage by then US Army Major Clay Lyle, regarding the effect of HEAT on unarmored wheeled vehicles:

> The Iraqis began using dump trucks to bring groups of fifteen to twenty-five reinforcements at a time to the battle … Eight dump trucks attempted to bring soldiers to the battlefield, and all eight were destroyed; 120mm HEAT from the tanks was completely destroying the trucks and all their occupants in one large blast. The Iraqis were blown from the dump trucks like popcorn out of a hot kettle.

A HEAT round utilizes a shaped charged explosive in a steel case with a copper-lined cavity at the front of the detonating charge. This hollow space directs the force of the explosion forward towards the target. Protruding in front of the projectile is a steel spike, or standoff tube, that allows the explosives to be detonated just before the body of the projectile impacts the target. This feature gives the exploding shaped charge time to form into a jet of superheated metal and gas capable of burning through target armor.

Another 120mm HEAT round used by the M1A1 and M1A2 series tanks is the M830A1 HEAT-MP-T round. Despite a designation that makes the M830A1 sound like a product-improved version of the M830 HEAT-MP-T, they are completely different rounds. In tank fire commands, the M830 HEAT-MP-T is referred to as 'MPAT.'

The M830A1 HEAT-MP-T (MPAT) round is also saboted requiring tank crews using the ammunition to be aware of any friendly personnel in the line of fire. It also has a proximity fuze, allowing it be fired at either ground targets or aerial targets, such as hovering or slow-moving helicopter gunships, which are a very potent threat to all tanks on the modern battlefield.

Main Gun Rounds in Combat

According to an online document from the US Army Abrams tank program manager titled *Lessons Learned Operation Iraqi Freedom 2003* appears this extract regarding the types of main gun rounds employed: 'Overall, very little SABOT was used. Devastating effects when used. HEAT and MPAT ended up being the preferred main gun round, effective against buildings and bunkers.' That extract is seconded in another post-Operation Iraqi Freedom report from the 3rd Infantry Division (Mechanized):

> Tanks were effective against most enemy direct fire targets. However, they were overkill in many cases. The armor-piercing fin stabilized discarding sabot APFSDS/high explosive antitank (HEAT) mix was adjusted in the ammo basic load (ABL) at the onset and was continually refined as the target array of trucks, BMPs, and bunkers became better known. Towards the end of offensive operations, ABLs consisted of about 60 percent HEAT and 40 percent APFSDS.

From the same 3rd Infantry Division (Mechanized) After-Action Report on Operation Iraqi Freedom appears this quote on the MPAT main gun rounds:

> We used the multipurpose antitank round (MPAT) in air mode to do the same thing [antipersonnel] but it would be more effective if it did not rubble everything around it. We could have fought the war with only the MPAT though.

The M830A1 HEAT-MP-T (MPAT) round is supposed to have an antipersonnel role. However, some Marine tankers in Iraq found that it was not as effective as hoped for when engaging enemy combatants. This is due to the physical characteristics of all shaped charged warheads, which tend to be very concentrated in their destructive effects.

The US Army no longer had a gun-armed combat engineer tank in service at the time of Operation Iraqi Freedom to destroy obstacles such as concrete bunkers and walls. It therefore went ahead and modified the MPAT round to perform the same role. In its modified form it was labeled the High-Explosive Obstacle-Reducing Tracer (HE-OR-T). It was first employed during the Iraqi Insurgency and quickly became the preferred main gun round for urban combat, as it proved much more effective in that environment than M830 HEAT-MP-T (HEAT) or M830A1 HEAT-MP-T (MPAT). Some Abrams tankers who served in Iraq would have preferred having all HE-OR-T rounds on their tanks.

Main Gun Round Secondary Effects

From an article that appeared in the November-December 2004 issue of *Armor* magazine titled *Sadr City: The Armor Pure Assault in Urban Terrain*, by then US Army Captain (now Lieutenant Colonel) John C. Moore comes this passage regarding the effectiveness of 120mm main guns firing any type of main gun round in an urban combat environment:

> Mahdi army elements are intimidated by 120mm main gun engagements. As soon as we began destroying the enemy with 120mm main guns, the enemy broke and ran. These engagements were often at short range where the concussive effect of the cannon was lethal, even if the enemy was not directly hit by the rounds …

The powerful concussive shock wave generated by the 120mm main gun rounds from the Abrams tank mandates that the vehicle commander and loader's heads must be below hatch level when a main gun round is fired. All friendly non-tank personnel should be at least 55 yards away from a firing tank to prevent a chance of injury. Abrams tankers who served in Iraq often recount the number of broken windows in urban areas generated upon firing their main guns.

Other Main Gun Rounds

Another main gun round now found in the ammunition racks of the M1A1 and M1A2 tanks is the relatively short-range (500 yards), blunt nose anti-personnel round officially known as the Tank, Cartridge, 120mm Canister M1028. It entered service in 2004. In tank fire commands the M1028 is referred to as 'Canister,' or its nickname 'CAN.' It sends out 1,100 small tungsten balls in a shotgun-like pattern. According to the designer and builder of the M1028 round, it 'can be used to clear enemy dismounts, break up hasty ambush sites in urban areas, clear defiles, stop infantry attack and counterattacks, and support friendly infantry assaults by providing cover-by-fire.'

The eventual replacement for the M830A1 HEAT-MP-T (HEAT) and the M1028 antipersonnel rounds on M1A1 and M1A2 tanks is labeled the M1069 Advance Multi-Purpose (AMP) round. What the new 120mm M1069 AMP round offers to tank crews appears in a January–February 2014 issue of *Armor* magazine in an article titled *The Armored Brigade Combat Team 2014–2024: Improving Abrams Lethality*, by Major Robert Brown:

> The advanced multi-purpose (AMP) round is a line-of-sight munition with three modes of operation: point detonates, delay, and air burst. The essential capability required in urban environments allows the tank to defeat AT [antitank] guided-missile teams at ranges of 50 to 2,000 meters [54 yards to 2,187 yards] with a precision lethal airburst. The point-detonates and delay modes allow for obstacle reduction (OR), bunker defeat and a wall breach capability for dismounted infantry …

In addition to the combat rounds for the M1A1 and M1A2 tanks there are comparable training rounds that match the flight path of the combat rounds, but not their maximum range, so as to remain within the physical confines of existing tank ranges in the United States and overseas.

The Marines Go Their Own Way

The US Marine Corps decided to take a different path in 2013 with its choice of main gun rounds for their M1A1 Common tank inventory. They bought a large number of a new main gun rounds, referred to as the 120mm Multi-Purpose High Explosive (MP-HE), to replace some of the existing main gun rounds. The new round was experimentally tested in Afghanistan beginning in 2011 and quickly proved its worth. It comes with three fuze settings: point detonation, delay, and airburst. An Office of Naval Research fact sheet describes the capabilities of the MP-HE round:

> The point detonating fuse will allow for the destruction of light armored vehicles and bunkers at extended ranges due to fin stabilization of the round. The delay mode will provide enhanced breaching capabilities for dynamic entry by dismounted infantry, and provides greater internal blast fragmentation within a building to destroy concealed enemy targets. Finally, the potential air burst capability of the round duplicates the effects of the canister round [M1028], but increases the maximum effective range from 300 meters to 5,000 meters to adequately engage [anti-tank guided missile] teams.

What Happens When the Main Gun is Fired

When an M1A1 gunner, or M1A2 gunner or tank commander identifies a suitable target and fires the main gun an electrical current flows through the firing pin, setting off the electrical primer of the cartridge case in the chamber, just in front of the gun's breechblock. An explosion (really a rapid burning of the propellant within the cartridge case) then takes place. In fewer than 16.4 feet and in less than 0.01 of a second, the projectile accelerates to Mach 5 (3,580 mph). The pressure in the chamber approaches 100,000 pounds per square inch for some types of main gun ammunition.

The sudden expansion of propellant gases and forward movement of the projectile produces an opposite rearward action of the gun tube, a maximum of 13 inches. This force is referred to as recoil and is very powerful in large-caliber guns. The M1A1 and M1A2 tank recoil mechanisms act as a cushion between the gun tube and gun mount, allowing the gun tube to move rearward while the gun-mount and the vehicle remain stationary. A counter-recoil system returns the gun tube to its original firing position (referred to as battery) and holds it there until fired again. In the process of returning to battery the breech block of the main gun opens and ejects the case base of the main gun round.

Bore Evacuator

Midway down the barrel of the M1A1 and M1A2 tank 120mm main gun is a device referred to as a bore evacuator, also seen on the M1 and IPM1. It is a fiberglass chamber set over the gun tube, which covers a number of small holes, which are inclined so that their inner [bore] ends are closer to the muzzle than the chamber where the round is inserted. Its job is to prevent the poisonous propellant gases generated by firing the tank's main gun from entering the turret when the breech-block opens upon counter-recoil.

When a projectile travels past the barrel holes, hidden under the bore evacuator, some of the propellant gases hurling the projectile down the barrel vent through the holes and pressurize the bore evacuator chamber. Upon the projectile exiting the muzzle end of a tank gun, the bore returns to near-atmospheric pressure. The propellant gases then escape back into the tank gun barrel through the holes. Due to the inclination of the holes the propellant gases vent through the muzzle of the main gun, which creates the massive fireball seen in pictures. As the propellant gas from the bore evacuator exits the muzzle it leaves a slightly lower pressure between the bore evacuator holes and the chamber where the round was inserted. Once the breech block opens, any remaining propellant gases remaining are therefore vented towards the muzzle end of the main gun barrel.

Ammunition Storage and Handling

The M1A1 tank series has storage space for 40 main gun rounds and the M1A2 tank series has room for 42 main gun rounds, with 36 in the armored rear turret bustle and 6 inside the vehicle's hull. All the rounds are stored behind blast-proof doors, within compartments that will vent outward and away from the vehicle if they are detonated. The only time the vehicle's crew becomes vulnerable to an onboard ammunition explosion is during the brief periods that the loader is withdrawing a main gun round from behind one of the blast proof doors.

The US Army standard for loading the 120mm main gun on all models of the M1A1 and M1A2 tanks is 5 seconds. A very quick loader can do it in under 4 seconds. Shoulder, knee, and foot guards protect the loader from the dangers of the recoiling breech and ejected case base. All of these safety barriers fold out of the way when not being used.

Fire-Control

Having the most powerful tank gun and the deadliest ammunition means little without the ability of a tank crew to effectively acquire, track, and engage enemy targets under a variety of battlefield conditions. Already mentioned is the hunter-killer sighting arrangement that first appeared on the M1A2 series, which is referred to as the CITV. The main optical sighting system on the M1A2 series remains the GPS.

The GPS on the M1 through M1A2 tanks is linked to the main gun in both elevation and azimuth and is fully stabilized allowing it to engage targets while on the move. The exterior portion of the GPS is housed in an armored box, nicknamed the 'doghouse,' which protrudes through the roof of the M1 through M1A2 tank turrets, in front of the tank commander's cupola. In case of GPS failure, the gunner can use a backup optical sight on the right side of the main gun, labeled the Gunner's Auxiliary Sight (GAS).

A key part of the GPS is the Thermal Imaging Sight (TIS), there also being a TIS in the CITV on the M1A2 series tanks. The TIS detects the small outside temperature differences between objects, making warm-bodied soldiers and vehicles stand out in the TIS in contrast to their colder surroundings. Vehicles have internal temperature variations that form visible patterns. In the TIS, the shapes of the hottest vehicle components, such as engines and exhausts, appear the brightest. Objects with a moderate temperature, such as warm tank tracks, appear slightly dimmer. Objects with a cooler temperature, such as the hull of a tank, appear black. A well trained gunner or tank commander on an M1A1 or M1A2 series tank can use the TIS to determine from engine and exhaust cues whether the vehicle they are observing is a tank or truck, whether it's a front engine or rear engine vehicle, and whether it is making evasive maneuvers.

Besides its ability to see at night, the TIS on the M1 through M1A2 series tanks can peer through a variety of natural and artificial visual conditions. During the US Army and US Marine Corps' time in Iraq, the insurgents aware of the TIS would often build fires to hide behind as they engaged the American tanks, knowing the TIS would have a harder time seeing them. The insurgents that did not grasp this tactic did not last long according to American tankers who served during the Iraqi Insurgency.

Another important part of the GPS is the tank's laser rangefinder. It is labeled the LRF and is integrated into the gunner's GPS or tank commander's CITV for relative target ranging. Range calculations are displayed superimposed on the optical sight pictures. If the LRF goes down the gunner and tank commander can set the range data for their gun manually.

To assist the gunner and tank commander on the M1 through M1A2 series tanks there is an onboard digital fire-control computer. It compensates for various fire-control issues that can throw off the accuracy of a tank main gun when firing, such as thermal bending caused by uneven heat distribution along the gun tube. It also accounts for propellant temperature, gun tube wear, air temperature, wind speed and direction, cant, and angle of sight. In addition, it compensates for the motion of a moving target and/or the moving Abrams tank.

To keep all the electronic devices that make up the M1A2 series tank fire-control systems cool, the vehicle has an onboard air-conditioning system. It is referred to as the thermal management system (TMS) and consists of two major components: the vapor compression system unit (VCSU) located in the left rear section of the tank's turret bustle, and the air-handling unit (AHU) located in the front of the GPS.

In this picture we see the bore end of a 120mm main gun on an Australian Army M1A1 SA tank. The device on the top end of the barrel's bore is referred to as the Muzzle Reference Sensor (MRS). It provides the gunner with a reference point to determine gun tube bend for manual input to the tank ballistic fire-control computer. (*DOD*)

This line drawing of the 120mm main gun fitted to the M1A1 and M1A2 tanks shows some of its key features. A license-built near copy of a German-designed weapon from Rheinmetall, the American-built version of the gun has fewer parts. The barrel itself is a one-piece high-strength steel forging. (*DOD*)

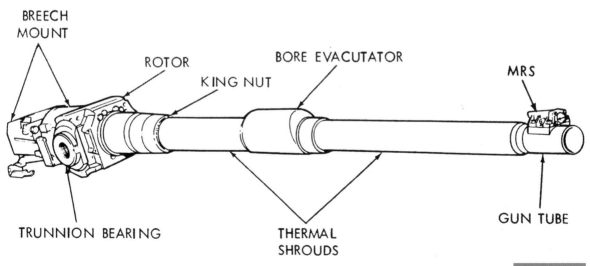

145

(*Left*) Shown here on display is an inert example of a 120mm training round of the M829 series of Armor-Piercing, Fin-Stabilized, Discarding Sabot, Tracer (APFSDS T). To the left of the training round is the stub, referred to as the case base, which remains from the combustible cartridge case upon being ejected from the main gun breech during counter-recoil. The vertical part of the stub is the primer element. (*Michael Green*)

(*Opposite page*) A young US Army tanker holds a training counterpart of the M829 series Armor-Piercing, Fin-Stabilized, Discarding Sabot, Tracer (APFSDS-T) main gun round. A problem with the M829 series APFSDS-T main gun round is that you do not want to fire it over the heads of friendly troops as its discarding sabot portions can kill or seriously injure those who may be struck by them. The danger zone is 1,095 yards in front of the gun and 77 yards on either side of the gun when fired. (*Michael Green*)

(*Below*) A labeled line drawing shows an example of the M829 series Armor-Piercing, Fin-Stabilized, Discarding Sabot, Tracer (APFSDS-T) main gun round. It weighs approximately 46lbs and is about 39 inches long. The 120mm APFSDS-T main gun rounds employed on the M1A1 and M1A2 all have depleted uranium (DU) penetrators and contain no explosive element, using pure kinetic energy (KE) to inflict damage. (*DOD*)

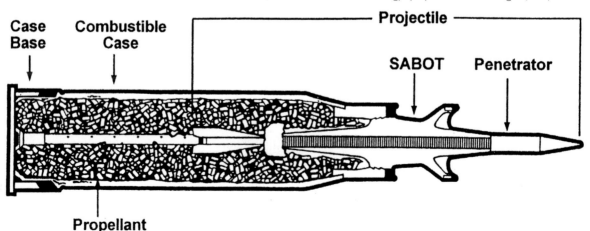

(Left) On the left is an inert training example of the 120mm M830 HEAT-MP-T main gun round and on the right an inert training example of the M829 series Armor-Piercing, Fin-Stabilized, Discarding Sabot, Tracer (APFSDS-T) main gun round. The M830 weighs approximately 53lbs and is about 38 inches long. *(Michael Green)*

(Below) A labeled line drawing shows some of the features of the 120mm M830 HEAT-MP-T main gun round. The fuse that sets off the high-explosive (HE) warhead on the projectile is referred to as a point-initiating, base-detonating and full frontal area impact switch. The fins on the end of the projectile reflect the switch from spin-stabilized tank ammunition to fin-stabilized tank ammunition for smoothbore tank barrels. *(DOD)*

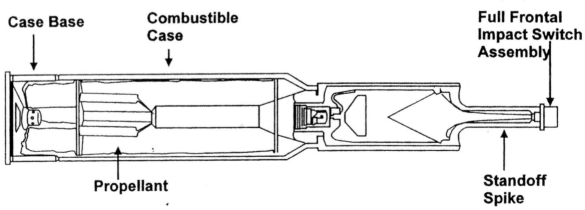

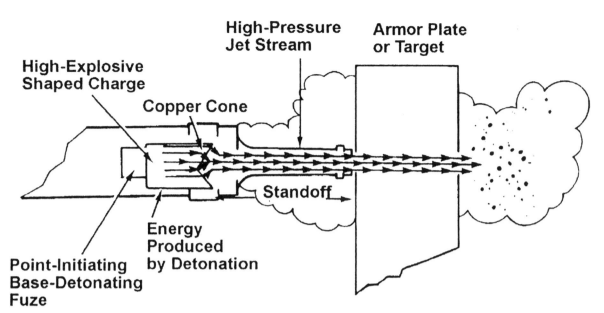

This line drawing shows how the 120mm M830 HEAT-MP-T main gun round functions. For the M1A1 and M1A2 tanks the M830 is considered to be a secondary armor-defeating main gun round. However, most modern tanks now feature armor that makes them immune on their frontal armor array from such rounds. However, it could still be effective on other portions of an enemy tank. (*DOD*)

A US Marine Corps tanker is shown holding a 120mm M830A1 HEAT-MP-T main gun round. It weighs approximately 50lbs and is about 39 inches in length. Like the 120mm M830 HEAT-MP-T main gun round it has a secondary armor-defeating role. It brought an important improvement to the capabilities of the M1A1 and M1A2 tanks as it could be employed to engage slow-moving or hovering helicopter gunships. (*DOD*)

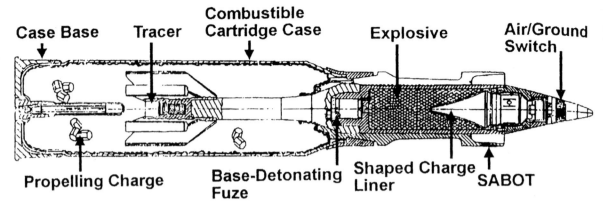

This line drawing shows the various features of the 120mm M830A1 HEAT-MP-T main gun round. Despite the ability to be set for air aerial targets, the M830A1 cannot be employed to engage fast moving helicopter gunships because the M1A1 and M1A2 tank fire-control systems are not fast enough to induce the correct lead angle for moving targets in the gunner's field of view. (DOD)

A US Marine Corps tanker is shown clutching a 120mm M830A1 HEAT-MP-T main gun round he has just removed from the rear turret main gun ammunition bustle behind his seat. The ready round blast door is still open in this picture. Abrams tanks have been struck in their rear turret bustles during combat resulting in the detonation of all of the main gun rounds stored within. Fortunately, the crews of the vehicles survived due to their blast proof doors and blow-off panels. (DOD)

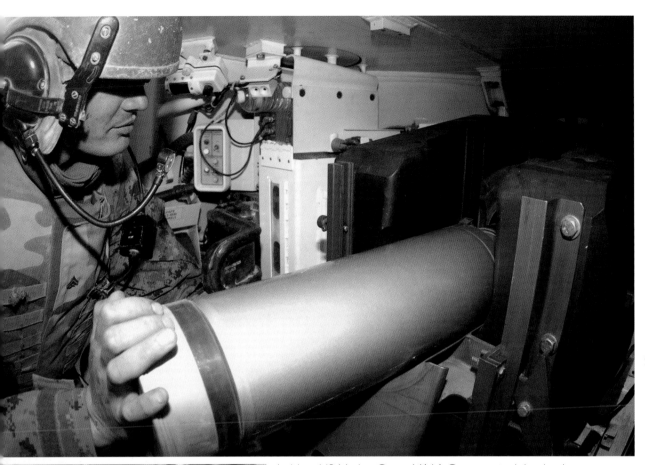

Inside a US Marine Corps M1A1 Common tank is a loader with a 120mm main gun round. Once the projectile end of the round is inserted into the open breech of the of the weapon, the loader makes a fist at the rear of the cartridge case and shoves it forcefully into the chamber just ahead of the breech-block, which quickly closes behind. (*DOD*)

On display at a military open house is a dummy training round that represents the appearance of the blunt-nosed M1028 anti-personnel canister round developed for use on the Abrams tanks armed with the 120mm main gun. The projectile portion of the round contains over 1,000 small tungsten pellets. It is intended to deal with large numbers of enemy ground personnel at ranges of under 500 yards. (*Norman A. Graf*)

(*Above*) A picture of a gunner of a US Army M1A2 tank, as indicated by the illuminated Gunner's Control and Display Panel (GCDP) just behind him. On American tanks since the Second World War, the gunner is typically the second most experienced man on a vehicle's crew and is normally seen as a tank commander in training. (*DOD*)

(*Opposite above*) On a firing range is a US Army M1A2 SEP V2 tank. American tankers are taught in training that when engaging multiple enemy targets on the battlefield, common sense is to be applied and they must take them under fire in the order of the threat they represent. Normally, if there is more than one threat, Abrams tankers are instructed to engage the closest one first. (*DOD*)

(*Opposite below*) The dust is shaken off the M1A2 SEP V2 tank as it fires its 120mm main gun. It normally falls to the Abrams tank commander to determine what type of main gun round is to be employed against a target, based on his knowledge of the target's vulnerabilities. When engaging enemy tanks it remains a truism that their weaker armor array is always on their sides and rear. (*DOD*)

Pictured during a training exercise is a US Marine Corps M1A1 Common tank that has just fired its 120mm main gun. When engaging a target on the firing range or in combat, Abrams tankers employ a standard terminology to speed up target engagement time. Unarmored vehicles are called 'trucks,' while armored personnel carriers are 'PCs.' Helicopters are called 'choppers' and enemy personnel are 'troops.' *(DOD)*

An M1A1 tank is shown at the moment of firing. The 120mm M830 HEAT-MP-T main gun round, like all HEAT rounds, is not as accurate as the M829 series Armor-Piercing, Fin-Stabilized, Discarding Sabot, Tracer (APFSDS-T) main gun round at ranges over 2,187 yards, as they have a slower muzzle velocity. However, using chemical energy (CE) instead of kinetic energy (KE) it remains effective against armored targets it impacts upon at any range. *(DOD)*

A less fun aspect of firing the main gun on an Abrams tank is cleaning the gun afterwards. Pictured is a US Marine Corps tanker swabbing the bore of the 120mm main gun on his M1A1 Common tank. There is an element of danger in cleaning the bore of the gun, or the bore evacuator, as traces of depleted uranium (DU) may still be present from the firing of DU main gun rounds. *(DOD)*

This picture was taken inside the turret of a US Army M1A2 SEP tank. Visible is the ready rack, located in the tank's left rear turret bustle, just behind the loader's seat. The ready rack holds 18 × 120mm main gun rounds. There are another 18 main rounds behind the tank commander's seat, on the right hand side of the rear turret bustle. *(Michael Green)*

(Left) This picture is looking into the hull of a US Marine Corps M1A1 Common tank with its turret having been removed. This would be the space for the tank's turret basket. Visible on the left side of the engine compartment bulkhead are the two blast proof doors, each of which has space for 3 main gun rounds. *(DOD)*

(Opposite page) In this picture we see the Gunner's Primary Sight (GPS) being removed from a US Army M1A1 tank with an overhead crane. Notice that the ballistic shield cover for the GPS has been unbolted and removed from the turret roof. On the top of the device are the independently stabilized head mirrors of the GPS that are protected by the ballistic shield cover. *(DOD)*

Seen in this close-up picture is the Gunner's Primary Sight (GPS) ballistic shield cover on an M1A1 series tank in Iraq. The armored doors are shown in the open position. Notice that the thick ballistic glass that fronts the gunner's upper optic sights, contained within the ballistic shield cover, has been penetrated. The ballistic shield cover on the M1 and M1IP differ in some construction details from those fitted to the M1A1 and M1A2 tank series. *(DOD)*

(*Above*) Here we see a crewmember of a US Army M1A2 SEP V2 bore-sighting the 120mm main gun. Firing a tank's main gun without accurate sight alignment is an exercise in frustration. For this reason, bore sighting is of the utmost importance in tank gunnery. The tanker in the picture is using a muzzle bore sight device to line up his gun tube with the tank's optical sighting devices. (*DOD*)

(*Opposite page*) Two US Army tankers are shown examining a brand new or rebuilt Gunner's Primary Sight (GPS) that has just been removed from its shipping container. The gunner on the M1A1 through M1A2 tank uses the GPS for both main gun and coaxial machine gun employment. Despite the many features included with the GPS, including a laser range finder (LRF), Abrams tank crews are taught visual ranging of targets as a backup. (*DOD*)

Chapter Six

Specialized Vehicles and Accessory Kits

During the many decades in which the Abrams tank series has been in service with the American military, a very small number of specialized vehicles have been proposed, based on the tank's chassis. Some never went past the prototype stage, but a handful entered service. In addition, there have been a number of accessory kits developed for the Abrams tank series to improve their battlefield effectiveness.

Bridge Launchers

Among the specialized vehicles based on the Abrams chassis to enter American service is the M104 Wolverine Heavy Assault Bridge (HAB). It is a turret-less version of the M1A2 SEP V2 configured to carry into battle a German-designed and built portable bridge that can be employed to cover an opening of almost 80 feet. Once emplaced the bridge was designed to support a 70-ton vehicle crossing it at approximately 10 mph. The first production unit of the vehicle entered US Army service in 2003.

The US Army had originally planned on acquiring 465 units of the M104 Wolverine. They were intended to replace the bridge launchers in service that were based on the M60 series of main battle tanks, and referred to as the Armored Vehicle Launched Bridge (AVLB). However, as the heavy armored elements (sometimes referred to as the legacy forces) of the US Army continued to shrink in size, only 44 units were eventually acquired before the program was cancelled due to cost.

In 2012, the US Army awarded a contract to GDLS to construct two prototypes of a new Abrams-based portable bridge launcher referred to as the Joint Assault Bridge (JAB). They were to be delivered in 2014 for testing. The term 'joint' in the vehicle's designation means it is intended for use both by the US Army and US Marine Corps.

To keep the cost down for the JAB the shorter bridge launcher unit from the M60 AVLB is being used. It can only cover an opening 60 feet wide and support the weight of the M1A2 SEP V2 at extremely slow speed. However, due to funding constraints imposed by the American Congress it is unlikely JAB will ever be put into production.

Obstacle Clearing Vehicles

The M1 Panther II is a dedicated mine clearing vehicle that first appeared in US Army service in the early 1990s. It consists of a non-turreted IPM1 tank chassis fitted with mine rollers or a mine plough attached to the front of the vehicle. Reflecting the danger posed to the crews of all mine-clearing vehicles, the M1 Panther II can be operated by remote-control. It can also be operated manually by its two-man crew if the situation dictates. Only six have been constructed and they have seen employment in a number of locations, including Iraq.

In the early 1990s, the US Army had hoped to field a specialized variant of the Abrams, labeled the M1 Grizzly Combat Mobility Vehicle (CMV). It was intended to clear obstacles, both natural and man-made, for the advance of the gun-armed versions of the Abrams series of tanks. Only two prototypes were built before the program was cancelled in 2001, due to cost overruns.

The US Marine Corps saw promise in the CMV concept and resurrected a much modified version labeled the Assault Breacher Vehicle (ABV), with the first unit being delivered in 2008. The ABV was first used successfully by the US Marine Corps in 2009 to clear minefields installed by the Taliban, in Afghanistan. The promising results obtained by the ABV in Afghanistan resulted in the US Army adopting the vehicle.

A passage on the ABV appears in a publication titled *US Army Weapon Systems 2014–2015*. It provides this description of the vehicle:

> … a highly mobile and heavily armored minefield and complex obstacle breaching system. It consists of an M1A1 Abrams tank hull, a unique turret with two Linear Demolition Charge Systems (employing two Mine Clearing Line Charges (MICLIC)) and rockets, a Lane Marking System (LMS), Integrated Vision System, and a High Lift Adapter that interchangeably mounts a Full Width Mine Plow (FWMP) or a Combat Dozer Blade (CDB). The ABV requires a crew of two soldiers. It improves the mobility of Combat Engineers and has the speed to keep pace with the maneuver force.

Add-On Abrams Tank Kits

With the heavy damage inflicted on Abrams series tanks during the close-in urban fighting in Iraq during Operation Iraqi Freedom, there developed the impetus for the US Army to authorize the funding for a new program in 2004 to better equip the tank for a type of fighting it had never been designed to take part in. The program was referred to as the Tank Urban Survivability Kit (TUSK) and was publicly announced in early 2005.

In a *US Army News Service* article published on 9 March 2005 is this description by US Army Program Manager Lieutenant Colonel Michael Flanagan on why the new add-on armor kit was needed and what benefits it would provide the Abrams series of tanks: 'You have to remember, the tank was a Cold War design, aimed at threats that were always to its front. It's still the most survivable weapon in the arsenal from the front ... Today it's a 360-degree fight, and these systems are designed to improve survivability in the urban environment.'

In 2006, the US Army ordered 505 units of the TUSK kits from GDLS. The first Abrams series tank in Iraq fitted with the kit was in 2007. By the following year, most Abrams series tanks employed in country had the TUSK kits. However, some were never fitted with the kits due to shortages. Others had only certain components of the kits applied.

Reflecting the various versions of the Abrams series tanks that were deployed to Iraq during the Iraq Insurgency, there was a TUSK I kit for the M1A1 and a TUSK II kit for the M1A2 and M1A2 SEP, which included some additional items for fitting to those tanks upgraded with the TUSK I kits.

TUSK I

One of the most noticeable external features of the TUSK I kit fitted to the M1A1 and M1A2 is the use of reactive armor tiles, also commonly referred to as Explosive Reactive Armor (ERA), along the sides of the vehicle's hull in place of the original armored side skirts. As listed on the American manufacturer's website, the tiles are officially designated the XM-19 Abrams Reactive Armor Tile (ARAT) and first appeared in 2006.

The first public employment of ERA occurred during the Israeli military invasion of Lebanon in 1982, when they were photographed on the exterior of American-supplied M60 main battle tanks in service with the Israeli Army (IDF). The Israeli firm that made the ERA tiles assigned them the label 'Blazer Armor'. The original design work on the Blazer Armor was done by a West German researcher working together with the IDF.

ERA tiles consist of a steel box that contains a special plastic explosive fitted in-between two steel plates. How ERA tiles function when struck is explained in a 1988 article by then US Army Captain James M. Warford titled *Reactive Armor: New Life for Soviet Tanks*, which appeared in the January-February issue of *Armor* magazine:

> ... the plastic explosive inside the brick [tile] detonates. The force of this detonation is directed away from the brick's inner steel plate, and concentrates in the opposite direction of the attacking warhead. This explosion forces the HEAT-formed jet to malform and lose its energy so that the heavily-weakened jet is not capable of penetrating the tank's main armor.

The ARAT set applied under TUSK I consists of 64 tiles, with 32 on each side of the hull, divided into two rows of 16 each. According to the manufacturer's website the ARAT is insensitive to bullets and other types of small battlefield fragments and will only detonate when struck by a shaped charge warhead, found on the projectiles fired from shoulder-launched rocket propelled grenade launchers, such as the RPG-7.

Tusk II
Under TUSK II, the M1A2 received a new generation of ERA tiles in 2008, labeled the XM-32. They are referred to as ARAT II tiles by the manufacturer, the original version being relabeled the ARAT I. The ARAT II set consists of 78 tiles, with 32 on each side of the tank's hull, divided into two rows of 16 each. Unlike the ARAT I set, the ARAT II includes 14 additional tiles, 7 mounted on either side of the turret of the tank it is fitted to. Pictorial evidence seems to suggest that the turret tiles are not always fitted, for an undisclosed reason.

Unlike the large box XM-19 tiles, optimized for protection from horizontally-fired shoulder-launched rocket-propelled grenade launchers, those of the XM-32 look like roof tiles. When fitted to an M1A2 SEP they are slanted downward, as they are optimized to protect the tank from Improvised Explosive Devices (IEDs) that are configured to fire an Explosively Formed Penetrator (EFP). These shaped charge penetrators are currently employed on a number of American military weapons, including the TOW-2B antitank missile. Those used by the Iraqi insurgents were improvised and typically emplaced at ground level to fire upwards.

TUSK Crew Protection Upgrades
Due to the high demand from the tankers in Iraq, the initial production run of 130 units of the Loader's Armored Gun Shield (LAGS), one of the numerous components making up the complete TUSK kits, was rushed to theater in 2005, before the complete kit sets were assembled. When the complete TUSK I kits reached Iraq in 2007, the Loader's Armored Gun Shield was fitted with a thermal sight officially designated the Light Thermal Weapon Sight (LTWS).

With the introduction of the TUSK II kit there appeared a 360 degree open-topped armor shield for the M1A2 tank commanders, which included transparent armor to improve his visibility in combat when engaging the enemy. A much simpler open-topped shield arrangement had also been provided for the M1A1 series tanks under TUSK II. Externally, it appears the only TUSK feature adopted by the US Marine Corps in Iraq was the LAGS.

Due to the large-scale use of standard production antitank mines left over from the disbandment of Saddam Hussein's Army, and IEDs by the Iraqi insurgents, an important component that formed part of TUSK I was a V-shaped armored plate 200mm thick attached to the bottom hull plate of Abrams series tanks in-country. It weighed 2,998lbs.

Reflecting the widespread use of conventional land mines and IEDs by the Iraqi insurgents, part of the TUSK I included a new Mine Resistant Driver's Seat that minimized the effect of the blast and resulting shock wave on the driver. No longer was the driver's seat mounted to the floor, which would transmit a shock wave to the driver's body, but attached to the ceiling of the front hull compartment. In addition, the driver was provided with a four-point seat belt system to prevent him from being thrown around his compartment upon driving over a mine or IED.

TUSK Vision Upgrades

As part of TUSK I the drivers on the Abrams series tanks were provided with a thermal sight, which could be installed in their center periscope station if the situation dictated. It is labeled the Driver's Vision Enhancer (DVE).

The TUSK I kits for the M1A1 series tanks included a thermal sight for the tank commander's .50 caliber M2HB machine gun called the Remote Thermal Sight (RTS). The M1A2 did not require the RTS as it was already designed with one.

Miscellaneous TUSK Upgrades

To facilitate communication between the Abrams series tanks in Iraq and the dismounted infantry they often worked in conjunction with, the TUSK I kit included a Tank-Infantry Phone (TIP) placed at the right rear of the vertical engine compartment. It has an extension cable to allow an infantryman attempting to use the TIP to seek nearby cover if being fired upon.

To minimize the collateral damage caused by firing the 120mm main gun of the M1A1 and M1A2 series tanks deployed to Iraq when engaged in urban combat, the TUSK I featured a remote-controlled .50 caliber M2HB machine gun and a 200-round ammunition box attached to the mantlet of the tank. It is called the Counter Sniper/Anti-Material Mount (CS/AMM).

The .50 caliber M2HB machine that is the heart of the CS/AMM is fixed in position and is aimed by the gunner using his main gun controls. It can be fired separately from the main gun. From a 19 February 2008 article by Pfc. April Campbell, written for the *Army News Service*, comes this quote by 2nd Lieutenant Frank Simmons regarding the introduction of the CS/AMM in Iraq: 'We're still lethal at long ranges without destroying everything. The sniper rifle mitigates the collateral damage.'

TUSK Features Deleted

As part of the original TUSK I there had been a slat armor kit developed to protect the rear of the vehicle engine deck from strikes by rocket-propelled grenade launchers. At least one US Army armored unit equipped with the M1A2 SEP had them fitted for a time. For undisclosed reasons, maybe based on their limited use in the field, the slat armor kit was deemed impractical, and it was deleted from being part of the TUSK program.

Another feature proposed for the original TUSK program included a remote-control mount for the tank commander's .50 caliber M2HB machine gun on the M1A2. This feature did not make it into the final TUSK program and did not appear in service until the introduction of the M1A2 SEP V2, fitted with the CROWS. Another feature proposed for the TUSK program was a rear hull camera system. It, like the CROWS, did not make it into the TUSK program, but later appeared on the M1A2 SEP V2.

During the Cold War the US Army developed and fielded specialized Armored Vehicle Launcher Bridges (AVLBs). These were based on the M48 medium tank series and later on the M60 main battle tank series. With the advent of the faster and heavier Abrams tanks the US Army once again went about fielding a suitable AVLB, which in this case was based on the chassis of the Abrams tank seen here. It was designated the M104 Wolverine. (*GDLS*)

(Above) The M104 Wolverine used a German-designed and German-built two-piece portable bridge that deployed itself horizontally as seen here to minimize its signature on the battlefield. Previous American-designed and American-built two-piece portable bridges were raised vertically before being deployed, giving away their position in the process. *(GDLS)*

(Opposite above) The need for a dedicated Combat Engineer Vehicle (CEV) for use by combat engineers working under fire has always been well understood by the US Army. With the fielding of the Abrams tank there was an attempt to employ the chassis of the vehicle for a state-of-the-art engineer vehicle seen here. It was labeled the Grizzly Combat Mobility Vehicle (CMV). *(Christopher Vallier)*

(Opposite below) The Grizzly Combat Mobility Vehicle (CMV) had been designed with a number of very useful features to help in breaching enemy obstacles. One of these included the massive dozer blade seen at the front of the vehicle pictured. Despite its capabilities, the cost of the vehicle doomed the program, and it was cancelled. *(Christopher Vallier)*

(*Opposite above*) A US Marine Corps Assault Breacher Vehicle (ABV) is shown here fitted with a Full Width Mine Plow (FWMP), which uses a raking action to clear a safe path through minefields by bringing concealed or buried mines and Improvised Explosive Devices (IEDs) to the surface where they can be disposed of. The large skids that project out in front of the FWMP control the ploughing depth by remaining in contact with the ground. (*DOD*)

(*Opposite below*) Upon reaching a minefield the Assault Breacher Vehicle (ABV) fires (as shown here) one of its two Mine Clearing Charges (MICLIC), which are stored in armored protected compartments at the rear of the vehicle's turret. They are elevated prior to being employed. Once the MICLIC detonates a portion of a minefield, the ABV will use its Full Width Mine Plow (FWMP) to clear a tank-wide opening and mark it for follow-on vehicles with a Lane Marking System (LMS). (*DOD*)

(*Above*) In response to the ever present threat to its Abrams tank from Rocket Propelled Grenade (RPG) launchers and other weapons employed by the insurgents in Iraq the US Army developed add-on features to improve their survivability. These features were labeled the Tank Urban Survivability Kit (TUSK) 1 and later 2. The most noticeable component of TUSK 1 was the box-like Explosive Reactive Armor (ERA) tiles fitted to the hull sides of Abrams tanks, as with the M1A2 pictured. (*DOD*)

(Above) General Dynamics Land Systems (GDLS) was contracted to add the various components that made up the Tank Urban Survivability Kit (TUSK) 1 and 2 to the US Army's Abrams tanks. In this picture, we see an M1A1 Abrams series tank being fitted with Explosive Reactive Armor (ERA) tiles from TUSK 1 at a GDLS facility in Iraq. *(DOD)*

(Opposite above) The first use of Explosive Reactive Armor (ERA) tiles by the American military occurred during Operation Desert Storm in 1991. Rather than appearing on US Army tanks it showed up in theater on US Marine Corps M60A1 tanks as shown here. The ERA tiles were developed and paid for by the US Army who in the end decided not to use them on their own M60 series tanks. *(DOD)*

(Opposite below) Being directed onto a tank transporter in Iraq is a US Army M1A2 tank fitted with the Tank Urban Survivability Kit (TUSK) 1 Explosive Reactive Armor (ERA) tiles. A feature of TUSK II as seen on this tank is an opened-topped partial armored enclosure fitted to the Improved Commander Weapon Station (ICWS). It includes a front-mounted gun shield. *(DOD)*

(*Above*) Eventually the Iraqi insurgents began to employ Explosive Formed Penetrator Improvised Explosive Devices (EFP IEDs) against Abrams tanks. The US Army's answer was the fielding of a second-generation of ERA tiles, seen here mounted on both the turret and hull of this M1A2 tank, which was part of the Tank Urban Survivability Kit (TUSK) II. As the EFP IEDs were typically located at ground level and aimed upwards the ERA tiles of TUSK II were slanted downward, as is shown in the picture. (*GDLS*)

(*Opposite above*) Visible in this picture taken in Iraq is an M1A2 tank fitted with the opened-topped partial armored enclosure fitted to the Improved Commander Weapon Station (ICWS), which formed part of the Tank Urban Survivability Kit (TUSK) II. Yet it lacks the gun shield for the .50 caliber M2HB machine gun. This could have reflected the many shortages that faced the TUSK I and II programs and resulted in some tanks not having all the features authorized. (*DOD*)

(*Opposite below*) Because the .50 caliber M2HB machine gun is mounted so far forward to a fixed pedestal on the Commanders Weapon Station (CWS) of the M1A1 tank series, there was no gun shield developed for the weapon in either the Tank Urban Survivability Kit (TUSK) I or II. The Explosive Reactive Armor (ERA) tiles seen on the side hull of the vehicle pictured are labeled the ARAT II by the builder. (*DOD*)

(*Above*) This picture taken in Iraq shows a US Army M1A1 series tank fitted with the opened-topped partial armored enclosure fitted to the Improved Commander Weapon Station (ICWS), which formed part of the Tank Urban Survivability Kit (TUSK) II. The device bolted to the main gun mantlet is labeled the Counter Sniper/Anti Material Mount (CS/AMM). It mounts a .50 Caliber M2HB machine gun. (*DOD*)

(*Opposite above*) In this picture of a US Army M1A1 series tank in Iraq can be seen the .50 Caliber M2HB machine gun fitted to the Counter Sniper/Anti Material Mount (CS/AMM). The gun can be cocked and a rate of fire selected by the tank commander from within the armored confines of the turret. The ammunition container affixed to the CS/AMM holds 200 rounds. (*DOD*)

(*Opposite below*) Visible on the bottom hull of this US Army M1A1 series tank in Iraq is the front of the 200mm thick V-shaped belly armor kit. This addition reflects the fact that the Iraqi insurgents often buried their Improvised Explosive Devices (IEDs) in the roads and trails they determined would be used by Abrams tanks on their patrols. Notice the Blue Force Tracking (BFT) unit is affixed to the armored box that covers the upper portion of the Gunner's Primary Sight (GPS). (*DOD*)

(*Above*) The massive size of the M1A1 Abrams series tank parked next to an Iraqi Army T-72 tank is clearly apparent in this photograph. The T-72 tank weighs approximately 46 tons compared to the 70 tons of the Abrams tank and has an automatic loader for its 125mm main gun. The T-72 tank in the picture once belonged to the former army of Saddam Hussein, but was incorporated into service by the new Iraqi Army, trained under American military supervision. (*DOD*)

(*Opposite page*) The close-in combat that prevailed during the Iraqi Insurgency took a toll on the loaders of Abrams tanks, who often manned the exposed M240 7.62mm machine gun on the roofs of their tanks. A makeshift solution to the problem was the addition of sand bags around the loader's overhead hatch on Abrams tanks as is seen here. (*DOD*)

(*Below*) The Tank Urban Survivability Kit (TUSK) I solution for the high losses among Abrams loaders operating their roof-mounted M240 7.62mm machine gun in Iraq is seen here. It was labeled the Loader's Armor Gun Shield (LAGS). It appears in many of the previous pictures in this chapter. It included (as seen in this picture) an armored parapet to deflect small arms fire directed upwards at the exposed loader. (*Michael Green*)

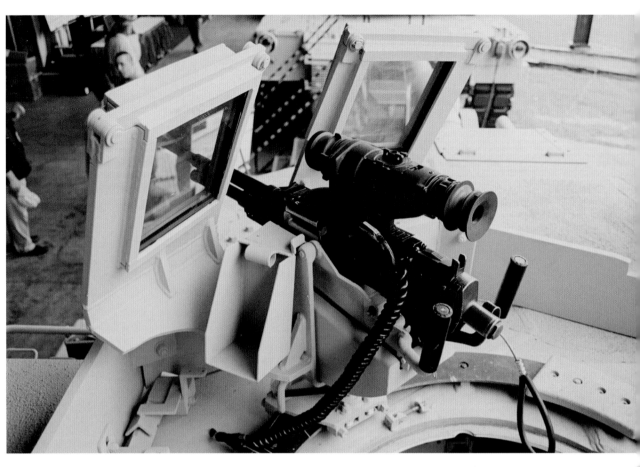

(Above) Seen here fitted to an M1A1 series tank in Iraq is a Full Width Mine Plow (FWMP). It consists of two blade units, one in front of each track, with attached skids to control the depth of the cut. The angled blade units direct uncovered mines and debris to the side of the tank as it moves forward to breach a path through a minefield. *(DOD)*

(Opposite above) A later feature of the Loader's Armor Gun Shield (LAGS) that appeared in Iraq is seen here mounted on a rail on top of the receiver of the M240 7.62mm machine gun and was designated the Light Thermal Weapon Sight (LTWS). In conjunction with the LTWS the loader, when operating the M240, can use a Helmet Mounted Display Sight (HMD). *(Michael Green)*

(Opposite below) Pictured in Iraq is an M1A1 series tank fitted with the Tank Mounted Mine Clearing Roller (TMMCR). It consists of three roller banks of five large steel discs each, with one located in front of each track, and another in the center in front. This was the same basic roller assembly as was developed for the M48 and M60 series tanks. *(DOD)*

(*Opposite above*) Mounted on the front hull of an M1A2 series tank in Iraq is a Full Width Mine Plow (FWMP). Located in between the two blade units is a chain and dog-bone like assembly employed to detonate any tilt-rod mines before they could go off under the hull of the tank. The FWMP can be attached or removed from a tank by four men in approximately sixty minutes when provided a suitable lifting device. (*DOD*)

(*Opposite below*) A US Marine Corps M1A1 Common tank in Iraq is fitted with the Combat Dozer Blade (CDB). According to the firm's website that designed and builds the device it can be used for a number of different purposes, including breaching defensive earthworks, preparing the ground for launching portable bridges, and for clearing urban roadblocks and rubble. (*DOD*)

(*Above*) Adopted by the US Marine Corps for some of its inventory of M1A1 Common tanks is the full-width Combat Dozer Blade (CDB) designed and built by the British firm of Pearson Engineering Limited. It also appears on the Assault Breacher Vehicle (ABV) employed by the US Marine Corps and US Army. The CDB weighs approximately 2 tons. (*DOD*)

The US Army has typically explored various experimental configurations of its in-service tanks to see if any might offer a significant advantage over the existing vehicles. One of these was an Abrams-based vehicle seen here labeled the Tank Test Bed (TTB). The 120mm main gun was mounted inside a small unmanned armored turret. The three-man crew was located side-by-side in the front hull. The early 1990s project never went anywhere. (Michael Green)

Another early 1990s experimental vehicle based on the Abrams tank is shown here. It was labeled the Component Advanced Technology Test Bed (CATTB). It was put together in anticipation of what would follow the planned M1A2 series tanks. Novel features of the CATTB included an autoloader, and the provision for a mounted 140mm main gun. Like the TTB the CATTB never went past the experimental stage and was later scrapped. (DOD)

Notes

Notes